Standards, Methods and Solutions of Metrology

Edited by Luigi Cocco

Published in London, United Kingdom

IntechOpen

Supporting open minds since 2005

Standards, Methods and Solutions of Metrology
http://dx.doi.org/10.5772/intechopen.77475
Edited by Luigi Cocco

Contributors
Denis Sullivan, Khalil Ebrahim Jasim, Imed Gargouri, Moncef Khadhraoui, Fatma Omrane, Oleh Velychko, Tetyana Gordiyenko, Chuxiong Hu

First published in London, United Kingdom, 2019 by IntechOpen
IntechOpen is the global imprint of INTECHOPEN LIMITED, registered in England and Wales, registration number: 11086078, The Shard, 25th floor, 32 London Bridge Street
London, SE19SG - United Kingdom
Printed in Croatia

British Library Cataloguing-in-Publication Data
A catalogue record for this book is available from the British Library

Additional hard and PDF copies can be obtained from orders@intechopen.com

Standards, Methods and Solutions of Metrology
Edited by Luigi Cocco
p. cm.
Print ISBN 978-1-78984-462-7
Online ISBN 978-1-78984-463-4
eBook (PDF) ISBN 978-1-83962-256-4

For EU product safety concerns:
IN TECH d.o.o., Prolaz Marije Krucifikse Kozulić 3, 51000 Rijeka, Croatia, info@intechopen.com or visit our website at intechopen.com.

Meet the editor

Dr. Luigi Cocco is a professional engineer with 15 years of experience in the automotive industry. He worked in electrical and electronic R&D and supply quality departments of racing/luxury/premium Italian vehicle manufacturers such as the Ferrari F1 Team, Automobili Lamborghini, and Alfa Romeo. Currently, he is ADAS System Responsible in Maserati S.p.A.

After his Master's degree in Telecommunication Engineering, he received his PhD in Information Engineering. He has published several papers and conference contributions on electronic measurements and digital signal processing. *Modern Metrology Concerns* and *New Trends and Developments in Metrology* are books edited by Luigi Cocco for IntechOpen respectively in 2012 and 2016.

Contents

Preface

The goal of acceptable quality, cost, and time is a decisive challenge in every engineering development process. To be familiar with metrology requires choosing the best combination of techniques, standards, and tools to control the project from advanced simulations to final performance measurements and periodic inspections. This book contains a cluster of chapters from international academic authors who provide a meticulous way to discover the impacts of metrology in both theoretical and application fields. The approach is to discuss the key aspects of a selection of untraditional metrological topics, covering the analysis procedures and set of solutions obtained from experimental studies. Each chapter is designed for both the engineering and academic communities and is partitioned as a scientific paper; a congruous list of references is available at the end of each section.

The volume has been divided into five chapters:

1. Metrological Traceability at Different Measurement Levels

Chapter 1 presents the basis of global metrological traceability and the standards of comparison of different regional metrology organizations. A procedure to evaluate specific interlaboratory results on national and international levels is described. This method can contribute the mutual recognition of measurement and testing results by different countries.

2. Self-Calibration of Precision XYθz Metrology Stages

Chapter 2 studies the on-axis calibration for precision XYθz metrology stages, and presents a holistic XYθz self-calibration approach. The proposed approach uses an artifact plate specially designed with XY grid mark lines and angular mark lines as a tool, to be measured by the XYθz metrology stages. Computer simulation is conducted and the designed artifact plate is illustrated to meet practical industrial requirements.

3. Third-Order Nonlinear Optical Properties of Quantum Dots

Chapter 3 introduces quantum dots (QDs), which are semiconducting nanocrystalline particles and attractive photonic media. Third-order nonlinear optical properties and a brief idea of the physics of QDs are discussed; the Z-scan technique and theoretical analysis adopted to obtain nonlinear parameters are detailed. Despite their size, QDs represent a good example of optical limiters with low threshold.

4. Analysis of Pulsating White Dwarf Star Light Curves

Chapter 4 is focused on analysis techniques for extracting the frequencies contained in the light curves of pulsating white dwarf stars. In several surface temperature regimes, these astronomical objects are unstable to gravity mode pulsations, which result in brightness variations corresponding to the periods of the excited modes.

Mode periods present in the light curve are detected by undertaking a Fourier analysis of the time series light curve.

5. **Biotoxicological Monitoring of Organic Solvents in the Tunisian Footwear Industry**

Chapter 5 is oriented to evaluate objectively the health effects of organic solvents that are widely used in the Tunisian footwear industry. The goal of the study is to identify analytical methods for exposure assessment of predominant solvents such as acetone, cyclohexane, hexane, methyl ethyl ketone, and toluene to arrange a process of occupational risk assessment via biotoxicological and airborne monitoring for solvents.

My appreciation goes to all authors for their valuable contributions. I would also like to thank my nephew Alessandro for his contagious energy.

Luigi Cocco, PhD
Maserati S.p.A., Italy

Chapter 1

Metrological Traceability at Different Measurement Levels

Oleh Velychko and Tetyana Gordiyenko

Abstract

The international agreements are the basis for establishing the global metrological traceability at different measurement levels. The concepts and concept relations around metrological traceability are presented. An important element of providing the metrological traceability is the evaluation of measurement uncertainty. The procedure of linking of key and supplementary comparison results is described. Linking of key and supplementary comparison results of the Regional Metrology Organization for some quantities according to the described procedure was presented. Results for all participants of presented key and supplementary comparisons are satisfactory for chi-square test and E_n number. The procedure of linking of key or supplementary comparison and national inter-laboratory comparison results is described. This procedure can be used for practical evaluation of specific inter-laboratory comparison results on a national level in different countries by means of laboratory results of the National Metrology Institute and Designated Institute. This procedure can contribute the mutual recognition of measurement and testing results by different countries. Linking of key comparison and inter-laboratory comparison results for some quantities according to the described procedure was presented. Results for all participants of presented key comparison and inter-laboratory comparison are satisfactory for chi-square test, E_n number, z scores and ζ scores.

Keywords: metrological traceability, measurement uncertainty, measurement standard, comparison, inter-laboratory comparison, National Metrology Institute, laboratory

1. Introduction

The Mutual Recognition Agreement (MRA) of the International Committee on Weights and Measures (CIPM) [1] and the MRA of the International Laboratory Accreditation Cooperation (ILAC) play an important role in overcoming technical barriers to international trade. CIPM MRA plays a key role in ensuring the international equivalence of national measurement standards of different countries. ILAC MRA plays a key role in ensuring international recognition of calibration results or test results in accredited calibration and testing laboratories. The main base of these agreements is special documents, guidelines, standards and recommendations [2].

National Metrology Institutes (NMIs) and Designated Institutes (DIs) play an important role in implementation of the CIPM MRA. They take an active part in organizing and conducting international comparisons of national standards.

IntechOpen

Consultative Committees (CCs) of CIPM and the International Bureau of Weights and Measures (BIPM) carry out key comparisons (KCs) of national standards in different fields of measurements. KCs are also being carried out by Regional Metrology Organizations (RMOs), which are equivalent to CC KCs. Only RMO makes supplementary comparisons (SCs) for those measurements that are not covered by KC CC or RMO. Results of all comparisons of standards are published in a special database KC (KCDB) of BIPM [3].

For CC KC and RMO KC, the reference value (RV) of KC and degree of equivalence (DoE) of national standards with corresponding uncertainty are established [4, 5]. DoE derived from an RMO KC has the same status as that derived from a CC KC. RMO SC has the same status as RMO KC. RMOs have a procedure to carry out comparisons, but only the Euro-Asian Cooperation of National Metrological Institutions (COOMET) has guidelines on comparison data evaluation [6, 7].

According to results obtained by the NMI or DI (NMI/DI) in conducted comparisons, Calibration and Measurement Capabilities (CMCs) of NMI/DI are being prepared [8, 9]. The internationally recognized NMI CMCs are those that are published to the KCDB of BIPM. Metrological traceability [10] is important for industrial metrology, because it allows you to compare measurement accuracy in accordance with a standardized procedure for assessing measurement uncertainty [11].

ILAC publication [12] established the need to ensure a continuous calibration chain to international or national standards as the main element for establishing metrological traceability. Important roles for the implementation of this requirement are calibration laboratories (CLs).

Inter-laboratory comparisons (ILCs) are a form of experimental verification of accredited calibration and test laboratories. They must meet the requirements of international standards ISO/IEC 17025 [13] and ISO/IEC 17043 [14]. Their main goal is to determine the technical competence of accredited laboratories for specific activities. The purpose of the ILC is to establish the inter-laboratory differences of their participants. Successful laboratory results in ILC confirm technical competence for certain types of measurements or testing.

Establishment of measurement traceability at the highest metrological level is carried out in accordance with procedures through international comparisons of the national standards of NMI/DI. Establishment of metrological traceability at lower measurement level is carried out in accordance with the calibration procedures of working standards by both NMIs/DIs and accredited CLs.

For the highest level of the metrological traceability, it is advisable to develop a methodology for linking of results of RMO SC to RMO KC, and RMO SC to other RMO SC. For lower level of the metrological traceability, it is advisable to develop a methodology for linking of results of the national ILC to RMO KC or RMO SC. These methodologies can be used for practical assessment of results of specific RMO KC/SC as an extension of the technical basis of confirmation of NMI/DI CMC or specific ILC and at the national level in different countries using the comparison results and CMC NMIs/DIs.

2. Bases of metrological traceability

The concept of metrological traceability is important for industrial metrology and is associated with such basic metrological concepts as measurement result, calibration chain, and measurement uncertainty [10]. A partial concept diagram around metrological traceability is shown in **Figure 1**.

The concept diagram demonstrates associative relations of metrological traceability with metrological traceability chain, measurement result, measurement

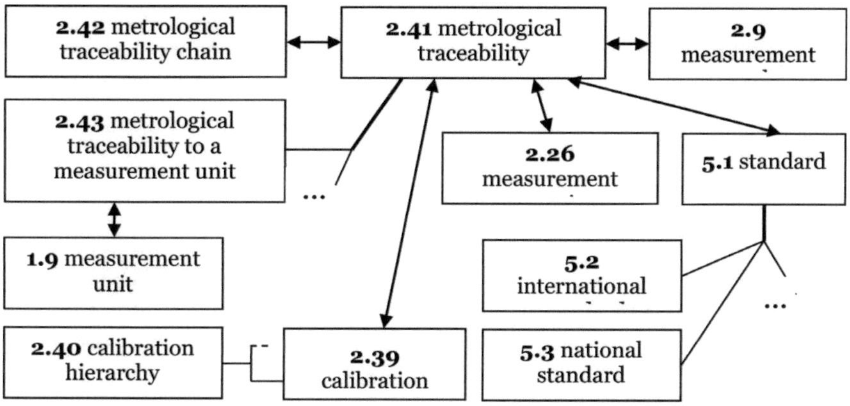

Figure 1.
Partial concept diagram around metrological traceability.

uncertainty, standard, and calibration. Hierarchical generic relations of metrological traceability with a measurement unit and of standard with international standard and national standard are established. Hierarchical partitive relation of calibration hierarchy with calibration is also established.

At the modern stage of development of the industrial metrology, the role of NMIs/DIs and CLs increases significantly. This is due to the need to ensure mutual recognition of measurement results in different countries. Global metrological traceability at different measurement levels [15] is provided by the CIPM MRA and ILAC MRA. These agreements set out the basic requirements for ensuring mutual recognition of both measurements and testing.

The general scheme of global metrological traceability at different measurement levels is presented in **Figure 2**.

International comparisons of national standards of NMIs/DIs are carried out as part of activities of the CIPM consultative committees (CCs) and technical committees of six RMOs. Results of these comparisons are technical basis for the preparation of NMI/DI CMC for publication in KCDB of BIPM. Accredited CLs and testing laboratories participate at the national level in the ILCs as part of activities of

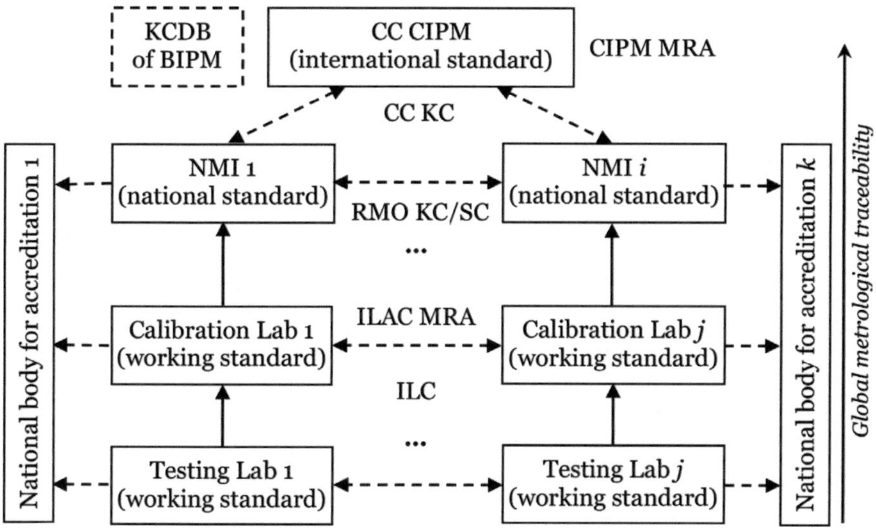

Figure 2.
The general scheme of global metrological traceability at different measurement levels.

national accreditation bodies. The calibration hierarchy is provided by calibration of the working standards and MIs: CLs—for testing laboratories; NMIs/DIs—for CLs.

3. The data evaluation of standard comparisons

The diagram of concept relations for standard comparisons is shown in **Figure 3**. Besides to KCs and SCs, pilot comparisons are also carried out, which all these comparisons can be bilateral. The organization of CC KCs and RMO KCs/SCs is the responsibility of pilot laboratory (PL) whose functions are performed by one of the selected NMI/DI [4, 6, 7]. The main responsibilities of PL include development of technical protocol of comparison, selection, and research of traveling standard, and the development of draft comparison reports. Coordination of the entire work of the PL as part of comparison is carried out by the contact person of PL.

The organizational scheme of standard comparisons is shown in **Figure 4**. NMI 1 is PL and is responsible for organizing the delivery of traveling standard to NMI participants. This scheme can be circular or radial. In the second case, it is better to provide research of drift of the traveling standard. The most commonly used is a mixed comparison scheme: after several NMI/DI participants of comparison, a traveling standard returns to PL for research of their drift.

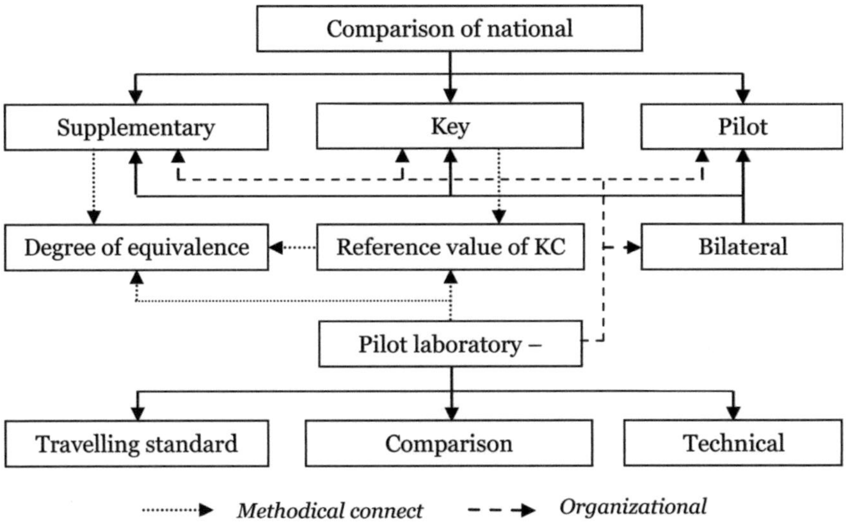

Figure 3.
The diagram of concept relations for standard comparisons.

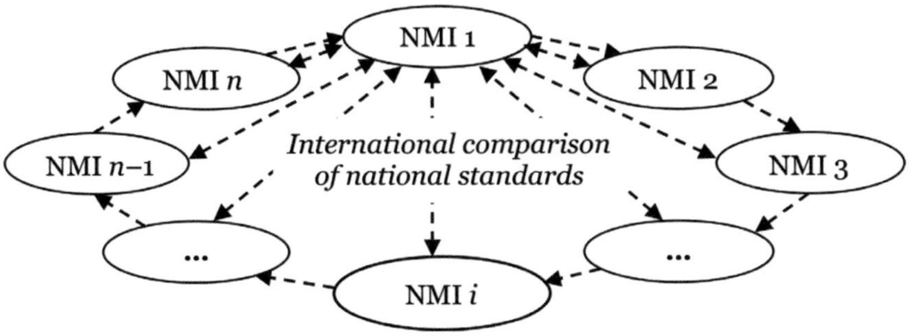

Figure 4.
The organizational scheme for standard comparisons.

RMO organizes KC with a number of joint NMI/DI participants with CC KC. This is necessary in order to link the results of the RMO KC with the results of the CC KC. For this purpose, equivalent technical protocols of both comparisons are used. The procedures for evaluating the data obtained at RMO KC are necessary to establish the DoE of national standards of NMI participants. PL calculates the KC RV and DoE for all NMI participants when preparing draft of comparison reports. The procedures used for evaluating RMO SC data are the same as for RMO KC. SC RMO complements KC CC or RM KC and is not second level comparison. RMO SC results are also published in KCDB of BIPM [16, 17].

RMO KC and RMO SC data evaluation usually includes determining the following characteristics: determining the RV comparison with the corresponding uncertainty, the DoE with corresponding uncertainties for each NMI/DI participant, and a pair DoE of i-th NMI/DI participant and j-th NMI/DI participant with corresponding uncertainties [6, 7]. RMO KC data evaluation includes the definition of such additional characteristics: converted KC data with corresponding uncertainties and DoE with corresponding uncertainties for each NMI/DI participant, except for linking NMI/DI.

The RMO KC/SC RV X_{RV} is calculated as the mean of NMI/DI participant results from RMO KC/SC data are given by

$$X_{RV} = \sum_{i=1}^{n} \frac{x_{NMIi}}{u^2(x_{NMIi})} \Big/ \sum_{i=1}^{n} \frac{1}{u^2(x_{NMIi})} \tag{1}$$

with the combined standard uncertainty

$$u^2(X_{RV}) = 1 \Big/ \sum_{i=1}^{n} \frac{1}{u^2(x_{NMIi})}, \tag{2}$$

where x_{NMIi} is the result for i-th NMI/DI participant in RMO KC/SC; $u(x_{NMIi})$ is corresponding standard uncertainty for i-th NMI/DI participant in RMO KC/SC; $i = 1, 2, \ldots, n$, n is the total number of NMI/DI participants of RMO KC/SC.

The DoE of i-th NMI/DI participant D_{NMIi} and corresponding combined standard uncertainty $u(D_{NMIi})$ are estimated as

$$D_{NMIi} = x_{NMIi} - X_{RV}, \tag{3}$$

$$u^2(D_{NMIi}) = u^2(x_{NMIi}) + u^2(X_{RV}). \tag{4}$$

Pairs of DoE of i-th NMI/DI participant and j-th NMI/DI participant D_{NMIij} of RMO KC/SC and corresponding combined standard uncertainty $u(D_{NMIij})$ are estimated as

$$D_{NMIij} = x_{NMIi} - x_{NMIj}, \tag{5}$$

$$u^2(D_{NMIij}) = u^2(x_{NMIi}) + u^2(x_{NMIj}). \tag{6}$$

On the basis of the measurement results of RMO KC/SC and corresponding combined standard uncertainties claimed by NMI/DI participants of RMO KC/SC, the chi-square test value is calculated [7].

$$\chi^2 = \sum_{i=1}^{n} \frac{D_{NMIi}^2}{u^2(x_{NMIi})}. \tag{7}$$

If the calculated chi-criterion value does not exceed the chi-square test critical value with the coverage level of 0.95 and freedom degrees of $n - 1$

$$\chi^2 < \chi^2_{0.95}(n-1), \tag{8}$$

then data can be acknowledged as consistent. This is the objective confirmation of declared uncertainties.

The NMI/DI participants of RMO KC/SC that provides maximum E_n number are determined [7].

$$\max_i E_n = \frac{|D_{NMIi}|}{2\sqrt{u^2(x_{NMIi}) - u^2(X_{RV})}}. \tag{9}$$

Then the data of NMI/DI participants with the largest value of E_n number are temporarily excluded from consideration, and the procedure for checking of consistency of the comparison data is repeated. Sequential data exclusion is repeated until the condition (8) is fulfilled.

The State Enterprise "Ukrmetrteststandard" (UMTS) was PL of several COOMET KCs and SCs in the field of electricity and magnetism (EM) in 2005–2018. UMTS as PL prepared and agreed with all NMI/DI participants draft reports on comparison COOMET.EM-K4, COOMET.EM-K5, COOMET.EM-K6.a, COOMET.EM-S2, COOMET.EM-S4, COOMET.EM-S13, COOMET.EM-S14, which comparison results are published in the KCDB of BIPM.

COOMET.EM-K4 comparison of national standards of a nominal capacitance of 10 pF at frequencies of 1000 and 1593 Hz was organized UMTS and carried out in 2005–2009. KV of COOMET.EM-K4 is $X_{KV} = -0.13$ μF/F at a frequency of 1000 Hz, and corresponding combined standard uncertainty is $u(X_{KV}) = 0.22$ μF/F ($k = 2$ for coverage level of 0.95). DoE for NMI/DI participants of COOMET.EM-K4 comparison for a nominal capacitance of 10 pF at a frequency of 1000 Hz [18] is shown in **Figure 5**.

Results of COOMET.EM-K4 comparison for a nominal capacitance of 10 pF at a frequency of 1000 Hz were checked for the fulfillment of the chi-square test. The obtained value of the chi-square test for all NMI/DI participants can be considered consistent, since the condition of expression (8) is satisfactory ($\chi^2 = 0.68 < \chi^2_{0.95}(n-1) = 1.15$). The same results of COOMET.EM-K4

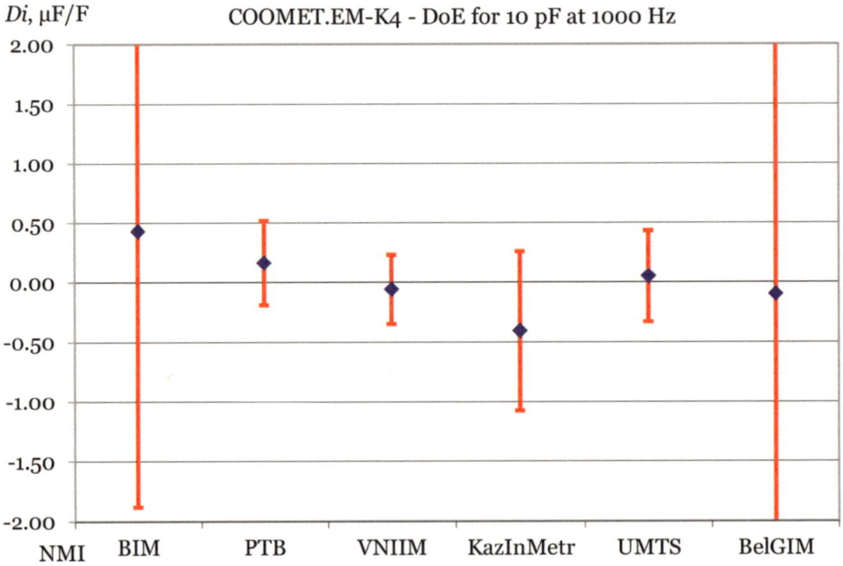

Figure 5.
DoE for NMI/DI participants of COOMET.EM-K4 comparison.

comparisons for all NMI/DI participants were checked for E_n number using Eq. (9). The resulting E_n number values for all NMI/DI participants do not exceed the value 1.0.

Results for the NMI/DI participants of COOMET.EM-K4 comparison are shown in **Table 1** for a nominal capacitance of 10 pF at a frequency of 1000 Hz.

COOMET.EM-K6.a comparison of AC voltage of 3 V at frequency of 20 kHz at frequencies of 1, 20, 100, and 1 MHz was organized UMTS and carried out in 2013–2014. KV of COOMET.EM-K6.a of AC/DC voltage transfer of AC voltage of 3 V at frequency of 20 kHz is $X_{KV} = -2.0$ μV/V, and corresponding combined standard uncertainty is $u(X_{KV}) = 1.9$ μV/V ($k = 2$ for coverage level of 0.95). DoE for NMI/DI participants of COOMET.EM-K6.a comparison for AC voltage of 3 V at frequency of 20 kHz [19] is shown in **Figure 6**.

Results of COOMET.EM-K6.a comparison of AC/DC voltage transfer of AC voltage of 3 V at a frequency of 20 kHz were checked for the fulfillment of the chi-criterion. The obtained value of the chi-square test for all NMI/DI participants can be considered consistent, since the condition of expression (8) is satisfied ($\chi^2 = 0.64 < \chi^2_{0.95}(n-1) = 0.71$ without INM data). The same results of COOMET.EM-K6.a comparisons for all NMI/DI participants were checked for E_n number using Eq. (9). The resulting E_n number values for all NMI/DI participants do not exceed the value 1.0.

Results for the NMI/DI participants of COOMET.EM-K6.a comparison are shown in **Table 2** for AC/DC voltage transfer of AC voltage of 3 V at a frequency of 20 kHz.

CMC [8] has three unambiguous characteristics: measurand, measurement range, and measurement uncertainty (generally given at a confidence level of 0.95).

NMI	BIM	PTB	VNIIM	KazInMetr	UMTS	BelGIM
D_{NMI}, μF/F	0.43	0.16	−0.06	−0.41	0.05	−0.10
$u(D_{NMI})$, μF/F	1.16	0.18	0.15	0.33	0.19	1.09
E_n	0.19	0.46	0.20	0.61	0.13	0.05

Table 1.
Results for NMI/DI participants of COOMET.EM-K4 comparison.

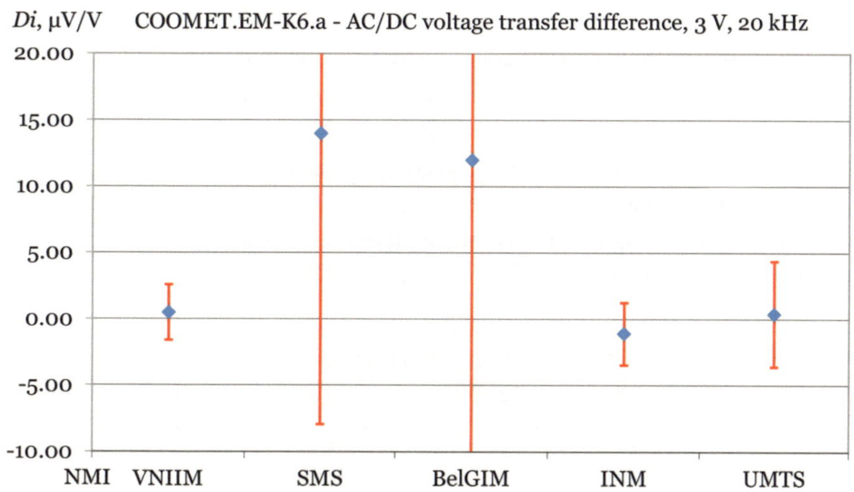

Figure 6.
DoE for NMI/DI participants of COOMET.EM-K6.a comparison.

NMI	VNIIM	SMS	BelGIM	INM	UMTS
D_{NMI}, μV/V	0.48	13.98	11.98	−1.12	0.38
$u(D_{NMI})$, μV/V	1.05	10.96	14.47	1.19	2.00
E_n	0.23	0.64	0.41	0.47	0.10

Table 2.
Results for NMI/DI participants of COOMET.EM-K6.a comparison.

They also contain a description of the used method or used measuring system, values of influence parameters, and any other relevant information. Normally for CMC, there are four ways in which a complete statement of uncertainty may be expressed: measurement range, equation, fixed measurand, and a matrix of measurement uncertainties.

CMC must be consistent with information from some or all of the following sources: results of KC and SC, knowledge of technical activities by other NMIs/DIs, including publications, other available knowledge and experience, etc. Results of RMO KCs/SCs are the ideal supporting evidence, but they can be used for fixed measurand only.

Methodologies for estimating the measurement uncertainty in a wide range of capacitance from 10 pF to 10 nF at frequencies of 1000 Hz and 1592 Hz and of inductance from 10 μH to 10 Hz at 1000 Hz are described in [20, 21], respectively. In these methodologies, requirements of both GUM [11] and regional recommendation [22] are used.

4. Linking procedures for international comparisons

Only CC KC results have a KC RV. Through joint NMI/DI participants, RMO KC must be linked to corresponding CC KC. The complete results of the linked RMO KC are presented in exactly the same form as the corresponding CC KC in KCDB of BIPM [4].

DoE of i-th NMI/DI participant of RMO KC is estimated as

$$d_{NMIi} = D_{NMIi} + \Delta, \tag{10}$$

where D_{NMIi} is result for NMI/DI participant from RMO KC only; d_{NMIi} is result for i-th NMI/DI participant which is linked to CC KC.

The correction factor for i-th linking NMI/DI is estimated as

$$\Delta_{iLink} = d_{iLink} - D_{iLink} \tag{11}$$

where d_{iLink} is result for i-th linking NMI/DI from CC KC; D_{iLink} is result for i-th linking NMI/DI from RMO KC.

The total correction factor Δ is then calculated as the weighted mean of the correction factor for linking NMI/DI participants, that is:

$$\Delta = \sum_{iLink}^{k} w_{iLink}\Delta_{iLink}, \tag{12}$$

$$w_{iLink} = \frac{s^2(\Delta)}{s^2(\Delta_{iLink})}, \tag{13}$$

$$s^2(\Delta) = 1/\sum_{iLink}^{k} \frac{1}{s^2(\Delta_{iLink})}. \tag{14}$$

The standard uncertainty $s(\Delta_{iLink})$ associated with Δ_{iLink} is calculated by the root-sum-square of the transfer standard uncertainty in CC KC: u_T is transfer standard uncertainty in RMO KC; $u(p_i)$ is standard uncertainty associated with the imperfect reproducibility of results of NMI_{iLink} in time period spanning two measurements; r_{iLink} is uncertainty associated with the imperfect reproducibility of measurement results of NMI_{iLink} in time period spanning its two measurements in CC KC and RMO KC; $i = 1, 2, \ldots, k$, k is total number of linking NMIs/DIs.

Table 3 lists the quantity values used in calculation linking total correction factor Δ and corresponding standard deviation $s(\Delta)$ for CCEM-K4 and COOMET.-EM-K4 comparisons for nominal capacitance 10 pF at a frequency of 1592 Hz [18].

The combined standard uncertainty is calculated as:

$$u^2(d_{NMIi}) = u^2(D_{NMIi}) + u^2(\Delta) = u^2(D_{NMIi}) + s^2(\Delta) + u^2(X_{RV}), \tag{15}$$

where $u(X_{RV})$ is combined standard uncertainty in CC KC RV.

The expanded uncertainty is $U(d_{NMIi}) = ku(d_{NMIi})$ which is chosen $k = 2$ for a coverage level of 0.95.

An example of linking of EUROMET.EM-K4, APMP.EM-K4.1, and COOMET.-EM-K4 results to the CCEM-K4 results for nominal capacitance of 10 pF at frequency 1592 Hz [18, 23, 24] is shown in **Figure** 7. When linking results of those comparisons, the presented linking procedure was used.

Results of EUROMET.EM-S26 comparison have been linked to EUROMET.EM-S20 comparison (two RMO SCs for an inductance of 100 mH at frequency

Linking NMI	d_{iLink}	D_{iLink}	Δ_{iLink}	u_T	$u(p_i)$	r_{iLink}	$s(\Delta_{iLink})$	w_{iLink}	Δ	$s(\Delta)$
VNIIM	−0.12	−0.10	−0.02	0.02	0.08	0.07	0.16	0.49	0.11	0.11
PTB	−0.00	−0.17	0.17	0.02	0.08	0.07	0.15	0.51		

Table 3.
CCEM-K4 and COOMET.EM-K4 data for linking NMIs, μF/F.

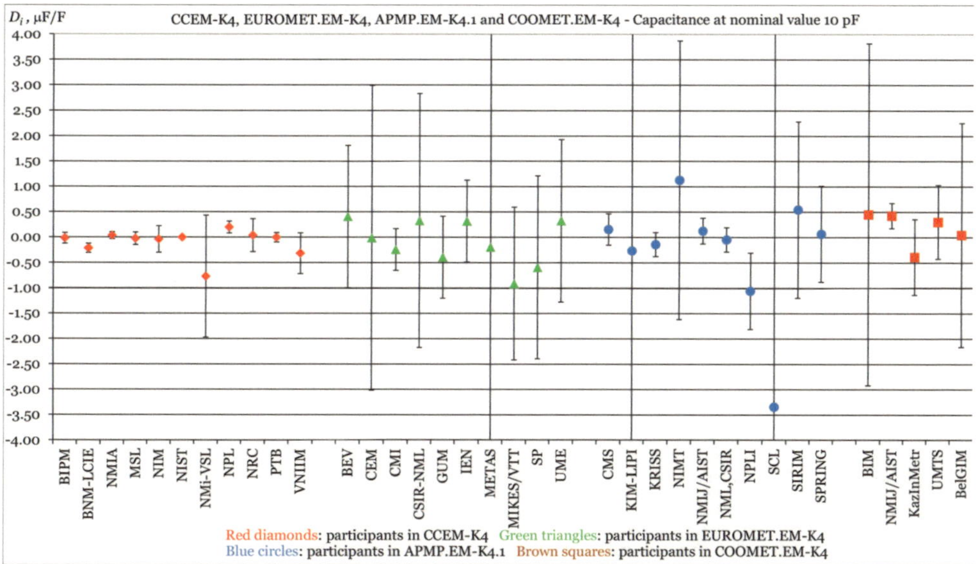

Figure 7.
Corrected DoE for participants of CCEM-K4, EUROMET.EM-K4, APMP.EM-K4.1, and COOMET.EM-K4 comparisons.

1000 Hz) with used special linking procedure [25] which is similar to the described linking procedure. Results of COOMET.EM-S2 comparison have been linked to EURAMET.EM-K5.1 comparison for electrical power [26]; results of COOMET.EM-S1 comparison have been linked to COOMET.EM-K6.a comparison of AC/DC voltage transfer difference [27] (RMO SC to RMO KC for similar values of physical quantities). When linking results of those comparisons, the described linking procedure was used.

Table 4 lists data for calculated total correction factors Δ and corresponding combined standard uncertainties $u(\Delta)$ for linking of COOMET.EM-S1 comparison results to COOMET.EM-K6.a comparison results for AC voltage of 3 V at frequencies of 1 kHz, 20 kHz, and 100 kHz [19], where X_{K6aKV} is COOMET.EM-K6.a RV; $u(X_{K6aKV})$ is combined standard uncertainty of COOMET.EM-K6.a RV.

Linked results of COOMET.EM-S1 (mark *) and COOMET.EM-K6.a comparison of AC/DC voltage transfer difference of AC voltage of 3 V at frequencies of 1, 20, and 100 kHz [27] are shown in **Figure 8**. When linking results of those comparisons, the presented linking procedure was used.

For consistency verification of results of COOMET.EM-K6.a and COOMET.EM-S1 comparisons, the value of chi-square test was calculated. The obtained value of chi-square test for all participants can be considered consistent: $\chi^2 = 0.58 < \chi^2_{0.95}(n-1) = 0.71$ (without VNIIM result) at frequency 1 kHz; $\chi^2 = 0.46 < \chi^2_{0.95}(n-1) = 0.71$ at frequency 20 kHz; and $\chi^2 = 0.49 < \chi^2_{0.95}(n-1) = 0.71$ (without VNIIM result) at frequency 100 kHz.

The maximum E_n number and declared uncertainties for DoE of NMI/DI participants of COOMET.EM-K6.a and COOMET.EM-S1 comparisons are judged as

Frequency	X_{K6aKV}	$u(X_{K6aKV})$	Δ	$u(\Delta)$
1 kHz	0.30	0.85	−0.60	1.15
20 kHz	−2.00	0.95	1.70	1.30
100 kHz	−6.80	1.70	5.60	1.85

Table 4.
Data for linking of COOMET.EM-S1 comparison results to COOMET.EM-K6.a comparison results, μV/V.

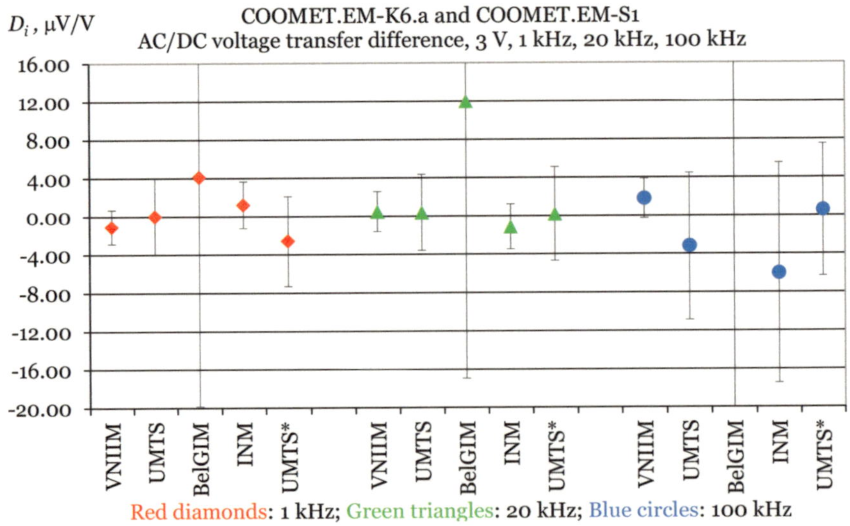

Figure 8.
Corrected DoE for participants of COOMET.EM-K6.a and COOMET.EM-S1 comparisons.

NMI	VNIIM	UMTS	BelGIM	INM	UMTS*
1 kHz					
D_{NMI}, μV/V	−1.10	0.00	4.10	1.20	−2.60
$u(D_{NMI})$, μV/V	0.90	2.02	11.97	1.22	2.35
E_n	0.28	0.00	0.17	0.28	0.44
20 kHz					
D_{NMI}, μV/V	0.48	0.38	11.98	−1.12	0.20
$u(D_{NMI})$, μV/V	1.06	2.00	14.47	1.19	2.45
E_n	0.11	0.07	0.41	0.26	0.03
100 kHz					
D_{NMI}, μV/V	1.81	−3.19	25.80	−5.99	0.60
$u(D_{NMI})$, μV/V	1.03	3.84	69.50	5.75	3.45
E_n	0.28	0.32	0.19	0.46	0.06

Table 5.
Results for NMI/DI participants of COOMET.EM-K6.a and COOMET.EM-S1.

confirmed by Eqs. (8) and (9) accordingly. Results for NMI/DI participants of COOMET.EM-K6.a and COOMET.EM-S1 comparisons are satisfactory (**Table 5**).

5. The data evaluation of national inter-laboratory comparisons

A number of studies are devoted to urgent questions of the data evaluation of ILC: the use of different methods for inconsistent data evaluation of ILC discussed in [28], suggested approaches to verifying the reliability of measurement results for CL participations of ILC [29], the application of z score test for performance evaluation of CLs recommended instead of E_n number since this number is not applicable due to the difficulty in determining the assigned value (AV) [30], algorithms for conducting ILC and obtaining precision data for CMC evaluation of laboratories are considered in [31–33], etc.

The general scheme of ILC is shown in **Figure 9**. Lab 1 is reference laboratories (RLs) of ILC. This scheme can be either circular or radial. Most often, a mixed

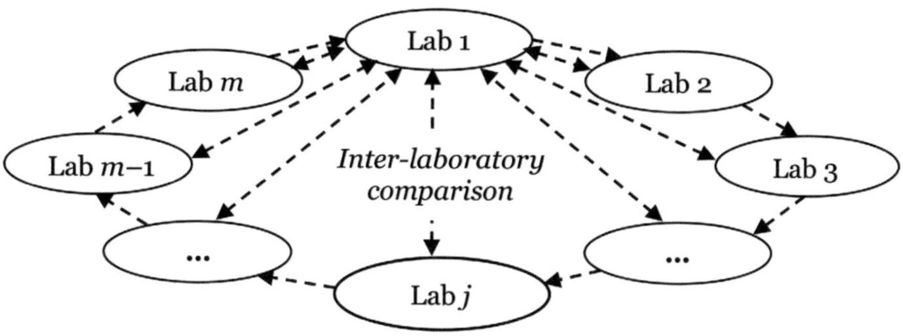

Figure 9.
The organizational scheme for ILCs.

scheme of ILCs is used: after several Lab participants, the traveling standard is returned to RL for research of its drift.

ILCs based on fundamental requirements: the repeatability and instability of traveling standard. Main steps common to nearly all ILCs are: the determination of AV, the calculation of performance statistics, the evaluation of performance, and the preliminary determination of ILC traveling standard stability [14].

The RL processes the data received from CL participants according to results of ILC for CL. Verification of ILC data is required for consistency. In the case of uncoordinated data, an analysis is conducted for the purpose of rejecting these data or for further harmonization by correction of the applied indicators. To verify the consistency of data, comparative analyses of the relevant criteria for performance statistics are carried out and the most effective for use in processing of the data is selected [14, 34].

There are various procedures available for the establishment of AV. These procedures involve the use of, in particular AVs—as determined by analysis, the measurement or standard comparison, traceable to a national or an international standard. The general algorithm for data evaluation of ILC is described in [35]. This algorithm allows RL to take into account all the reporting features of ILC.

The laboratory difference D_{labj} for j-th CL participant of ILC is calculated using Equation [14, 35, 36].

$$D_{labj} = x_{labj} - X_{AV}, \tag{16}$$

where x_{labj} is the measured value for i-th CL; X_{AV} is AV which is determined by RL.

The percent laboratory difference $D_{\%labj}$ for ILC is calculated using equation

$$D_{\%labj} = \left[D_{labj}/X_{AV}\right] \cdot 100. \tag{17}$$

The criteria for performance evaluation will be established after taking into account whether methods for evaluating the performance characteristics consider the main features, namely: the statistical determination of indicators, i.e. when the criteria must be suitable for each indicator; the compliance with the purpose, given criteria that take into account, for example, technical specifications for characteristics of method and recognized level of participant studies, etc. [14].

The most often to check consistency of ILC data that uses E_n number which is calculated using equation

$$E_n = D_{labj}/\sqrt{U^2\left(x_{labj}\right) - U^2(X_{AV})}, \tag{18}$$

where $U\left(x_{labj}\right)$ is the expanded uncertainty of a participant's result; $U(X_{AV})$ is the expanded uncertainty of RL's AV.

For an E_n number:

$|E_n| \leq 1.0$ indicates satisfactory performance;

$|E_n| > 1.0$ indicates unsatisfactory performance.

For checking consistency of ILC data, a z scores is also used, which is calculated by the equation

$$z = D_{labj}/\sigma, \tag{19}$$

where σ is the standard deviation for qualification assessment.

The value of σ can be calculated based on [14]: estimates from a statistical model (main model) or results of a precision experiment, estimates from previous ILC rounds or assumptions based on experience, results of participating laboratories, that is, normal or robust standard deviation, based on the results of ILC participating laboratories, etc.

For checking consistency of the ILC data, a ζ scores is used, which is calculated by the equation

$$\zeta = D_{labj} / \sqrt{u^2(x_{labj}) - u^2(X_{AV})} \quad (20)$$

where $u(x_{labj})$ is the combined standard uncertainty associated with result of the laboratory participating in the ILC; $u(X_{AV})$ is the combined standard uncertainty of ILC AV.

For a z scores and a ζ scores:

$|z| \leq 2.0$ and $|\zeta| \leq 2.0$ indicate a satisfactory performance characteristic and do not require adjustment or response measures;

$2.0\ |z| < 3.0$ and $2.0 < |\zeta| < 3.0$ indicate a dubious performance characteristic and require precautionary measures;

$|z| \geq 3.0$ and $|\zeta| \geq 3.0$ indicate an unsatisfactory performance characteristic and require adjustment or response measures.

Obvious blunders, such as those with incorrect units, decimal point errors, and results for a different ILC item will be removed from the data set and treated separately. These results will not be subject to outlier tests or robust statistical methods. If results are removed as outliers, they will be removed only for calculation of summary statistics. These results should still be evaluated within ILC scheme and be given the appropriate performance evaluation [35].

The value of expanded uncertainty $U(X_{AV})$ is estimated as

$$U(X_{AV}) = 2 \cdot \sqrt{u^2(x_{ref}) + u^2(x_{stab})}, \quad (21)$$

where $u(x_{ref})$ is the standard uncertainty obtained by calibrating traveling standard with a RL; $u(x_{stab})$ is the standard uncertainty from the instability of traveling standard during ILC period.

The value of standard uncertainty $u(x_{stab})$ is estimated as

$$u(x_{stab}) = \Delta X_{\max} / \sqrt{3}, \quad (22)$$

where $\Delta X_{\max}$ is the maximum change in nominal value of traveling standard during ILC period.

Linking the correspondingly expanded uncertainties of AV U_{AV} when RL of ILC are NMIs, accredited by CLs or accredited RLs that are not NMIs or accredited by CLs, is as follows [36]:

$$U_{AV\ NMI} < U_{AV\ CL} < U_{AV\ RL}, \quad (23)$$

that is, the most accurate ILCs are those that are performed by NMIs.

The value of the expanded uncertainty $U_{AV\ NMI}$ for a case where the NMI is RL can be derived from results of corresponding international comparisons of national standards in which the NMI participated. The value of the expanded uncertainty $U_{AV\ CL}$ for a case where CL is RL can be derived from corresponding calibration certificates for working standards issued by the NMI using CL in ILC. The value of the expanded uncertainty $U_{AV\ RL}$ for a case where an RL is an accredited provider can be obtained from corresponding calibration certificates for working standards issued by accredited CLs that use RLs in ILC.

An example of the laboratory difference D_{lab} of lab participants for national ILC of AC/DC voltage transfer difference of AC voltage of 3 V at a frequency of 20 kHz

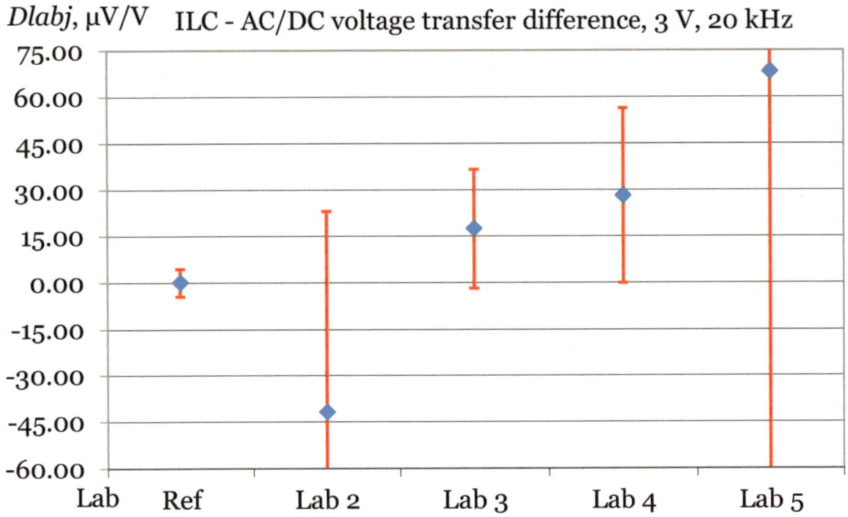

Figure 10.
Results of national ILC for AC/DC voltage transfer difference.

Lab	Ref	Lab 2	Lab 3	Lab 4	Lab 5
D_{lab}, µV/V	0.00	−42.00	17.40	28.10	68.20
$u(D_{lab})$, µV/V	2.25	32.50	9.60	14.10	1570.00
E_n	0.00	0.65	0.91	0.99	0.02
z	0.00	1.04	0.43	0.70	1.69
ζ	0.00	0.32	0.45	0.50	0.01

Table 6.
Results for all lab participants of ILC.

with respect to the AV with expanded uncertainty $U(D_{lab})$ [37] is shown in **Figure 10**.

For verification of consistency of the ILC results, the value of chi-square test was calculated. The obtained value of chi-square test for lab participants can be considered consistent: $\chi^2 = 2.52 < \chi^2_{0.95}(n-1) = 2.73$ (without Lab3 and Lab 4 results). Results for lab participants of ILC are satisfactory (**Table 6**).

6. Linking procedures for international comparisons and national interlaboratory comparisons

ILCs for CLs are carried out in different countries. To ensure the mutual recognition of calibration results, it is advisable to establish the relationship between these ILCs. To do this, NMI/DI results of international standard comparisons can be used. In this case, the DoE of NMI/DI standards and their uncertainty may be taken into account. Thus, it is possible to establish the metrological traceability of CL standards to corresponding national standards.

The organizational scheme of linking of international standard comparison and national ILC is shown in **Figure 11**. The Lab 1 is RL for ILC which is also i-th NMI for RMO KC/SC.

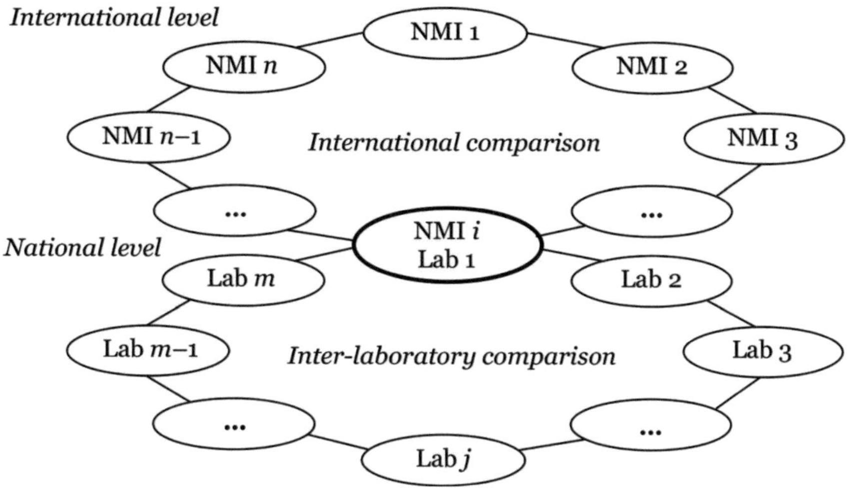

Figure 11.
The organizational scheme for linking of RMO KC/SC and national ILC.

In [38], the proposed procedure links RMO KC/SC and ILC results for CL. This procedure can be used for practical estimation of specific ILC results on a national level in different countries by means of NMIs/DIs results from RMO KC/SC.

The result of i-th NMI in some specific RMO KC/SC can be determined for linking in a specific ILC. Results of ILC will be expressed in relation to specific RMO KC/SC RV through linking laboratory—RL. For this purpose, the laboratory difference of ILC D_{labj} will be corrected by a correction factor d_{lab}, which is determined from the results of participant Lab 1 (RL) in RMO KC/SC and ILC (Lab 1 – NMI i):

$$d_{lab} = D_{NMIi} - D_{lab1} \tag{24}$$

with the combined standard uncertainty:

$$u^2(d_{lab}) = \left[u^2(D_{NMIi}) + u^2(D_{lab1})\right]/2. \tag{25}$$

The corrected DoE for j-th lab participant in ILC with respect to linking to RMO KC/SC RV is estimated as

$$D'_{labj} = D_{labj} + d_{lab} \tag{26}$$

with the combined standard uncertainty:

$$u^2\left(D'_{labj}\right) = u^2\left(D_{labj}\right) + u^2(d_{lab}) \tag{27}$$

The values of E_n number is determined by the equation

$$E_{nlabj} = \left|D'_{labj}\right|/U\left(D'_{labj}\right) \le 1.0. \tag{28}$$

The values of z scores is determined by the equation

$$z_{labj} = \left|D'_{labj}\right|/\sigma_{lab} < 2.0, \tag{29}$$

where σ_{lab} is the standard deviation, based on the results of ILC participating laboratories.

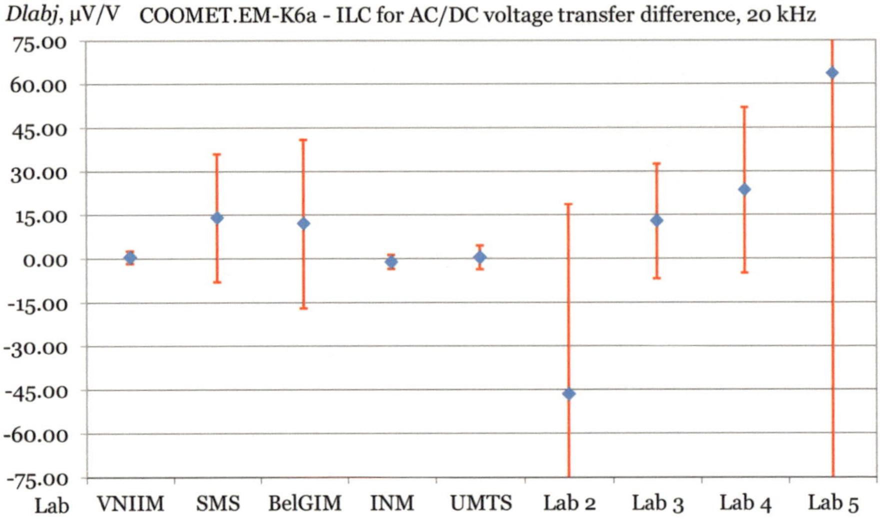

Figure 12.
The corrected laboratory difference for lab participants of national ILC for AC/DC voltage transfer standards with respect to linking to COOMET.EM-K6.a.

NMI-Lab	VNIIM	SMS	BelGIM	INM	UMTS	Lab 2	Lab 3	Lab 4	Lab 5
D'_{lab}, μV/V	0.50	14.00	12.00	−1.10	0.40	−46.60	12.80	23.50	63.60
$u(D'_{lab})$, μV/V	1.05	10.95	14.45	1.20	2.00	32.50	9.85	14.20	157.00
E_n	0.23	0.64	0.41	0.47	0.10	0.72	0.65	0.83	0.20
z	0.02	0.49	0.42	0.04	0.01	1.62	0.45	0.82	2.22
ζ	0.11	0.32	0.21	0.24	0.05	0.36	0.33	0.41	0.10

Table 7.
Results for all NMI/DI and lab participants.

The values of ζ scores is determined by the equation

$$\zeta_{labj} = \left|D'_{labj}\right| / u\left(D'_{labj}\right) < 2.0. \tag{30}$$

An example of the corrected laboratory difference D'_{lab} of lab participants for national ILC of AC/DC voltage transfer difference of AC voltage of 3 V at a frequency of 20 kHz with respect to linking to COOMET.EM-K6.a with expanded uncertainty [38] is shown in **Figure 12**. When linking results of those comparisons, the presented linking procedure was used.

For verification of consistency of COOMET.EM-K6.a and ILC results, the value of chi-square test was calculated. The obtained value of chi-square test for lab participants can be considered consistent: $\chi^2 = 0.71 \prec \chi^2_{0.95}(n-1) = 0.42$. Results for all NMI/DI and lab participants are satisfactory (**Table 7**).

7. Conclusions

CIPM MRA and ILAC MRA are the basis for establishing the global metrological traceability and play an important role in overcoming technical barriers to international trade. The calibration hierarchy and measurement uncertainty evaluation are important elements of providing metrological traceability. The general scheme of the global metrological traceability at different measurement levels is presented. NMIs/DIs and accredited CLs play an important role in establishing those traceability.

The organizational scheme for standard comparisons and RMO KC and RMO SC data evaluation procedure is presented. Results of data evaluation for COOMET.-EM-K4 and COOMET.EM-K6.a comparisons are indicated. Results of those comparisons were checked for the fulfillment of the chi-square test. The obtained values of the chi-square test for NMI/DI participants are satisfactory. Results for all NMI/DI participants of those comparisons for E_n number are also satisfactory.

The procedure of linking of RMO KC and RMO SC results is presented. Linking of COOMET.EM-S1 and COOMET.EM-K6.a comparison results of AC/DC voltage transfer difference at different frequencies is presented. The value of chi-criterion for linked comparison results was calculated. The obtained value of chi-square test for NMI/DI participants of those comparisons is satisfactory. Results for all NMI/DI participants of those comparisons for E_n number (from 0.03 to 0.46) are also satisfactory.

Results of linking of COOMET.EM-S1 and COOMET.EM-K6.a comparison results can also be used as the technical basis of confirming CMC NMIs/DIs. Such work can be done by PL of RMO KC or RMO SC, as well as by NMI/DI experts. The NMIs/DIs also must implement a full assessment of the uncertainty budget and the metrological traceability for validation of their CMCs in a wide range of used quantities.

The organizational scheme for ILs and ILC data evaluation procedure is presented. Results of data evaluation for ILC of AC/DC voltage transfer difference are indicated. Results of this comparison were checked for the fulfillment of the chi-square test. The obtained value of the chi-square test for laboratory participants is satisfactory. Results for all laboratory participants of this comparison for E_n number are also satisfactory.

The organizational scheme of linking of international standard comparison and national ILC is indicated. The procedure of linking of RMO KC or RMO SC and national ILC results is presented. This procedure can be used for practical estimation of results specific ILC on a national level by means of the results from NMI/DI laboratories. Linking of COOMET.EM-K6.a comparison and national ILC of AC/DC voltage transfer difference results was presented. The value of chi-square test was calculated and the obtained value of chi-square test for all participants can be considered consistent. Results for all participants of comparisons are satisfactory for E_n number (from 0.10 to 0.83), z scores (from 0.01 to 2.22), and ζ scores (from 0.05 to 0.41).

Results of this linking can be used also for different metrological areas as technical basis of confirming CMC accredited laboratories. Such work can be done by RL of the ILC, as well as by metrological experts. The RL of the ILC can also implement a full assessment of the uncertainty budget and the metrological traceability for validation of their CMCs in a wide range of used quantities.

Author details

Oleh Velychko[1]* and Tetyana Gordiyenko[2]

1 State Enterprise "Ukrmetrteststandard", Kyiv, Ukraine

2 Odesa State Academy of Technical Regulation and Quality, Odesa, Ukraine

*Address all correspondence to: velychko@ukrcsm.kiev.ua

References

[1] Text of the CIPM MRA [Internet]. 1999. Available from: https://www.bipm.org/utils/en/pdf/CIPM-MRA-2003.pdf [Accessed: 10-01-2019]

[2] Velychko O, Gordiyenko T. The implementation of general guides and standards on regional level in the field of metrology. Journal of Physics: Conference Series. 2010;**238**:012044:6. DOI: 10.1088/1742-6596/238/1/012044

[3] The BIPM key comparison database (KCDB) [Internet]. Available from: http://kcdb.bipm.org/ [Accessed: 10-01-2019]

[4] Measurement comparisons in the context of the CIPM MRA. CIPM MRA-D-05 [Internet]. 2016. Available from: https://www.bipm.org/utils/common/documents/CIPM-MRA/CIPM-MRA-D-05.pdf [Accessed: 10-01-2019]

[5] Velychko O, Gordiyenko T. The estimation of the measurement results with using statistical methods. Journal of Physics: Conference Series. 2015;**588**:012017:6. DOI: 10.1088/1742-6596/588/1/012017

[6] COOMET R/GM/14:2016. Guidelines for data evaluation of COOMET key comparison [Internet]. 2016. Available from: http://www.coomet.org/DB/isapi/cmt_docs/2016/5/2BMD1O.pdf [Accessed: 10-01-2019]

[7] COOMET R/GM/19:2016. Guideline on COOMET supplementary comparison evaluation [Internet]. 2016. Available from: http://www.coomet.org/DB/isapi/cmt_docs/2016/5/21XQGO.pdf [Accessed: 10-01-2019]

[8] Calibration and Measurement Capabilities in the context of the CIPM MRA. CIPM MRA-D-04 [Internet]. 2013. Available from: https://www.bipm.org/utils/common/documents/CIPM-MRA/CIPM-MRA-D-04.pdf [Accessed: 10-01-2019]

[9] Velichko ON. Calibration and measurement capabilities of metrological institutes: Features of preparation, examination, and publication. Measurement Techniques. 2010;**53**(6):721-726. DOI: 10.1007/s11018-010-9567-x

[10] International vocabulary of metrology—Basic and general concepts and associated terms (VIM). 3rd edition. JCGM 200 [Internet]. 2012. Available from: https://www.bipm.org/utils/common/documents/jcgm/JCGM_200_2012.pdf [Accessed: 10-01-2019]

[11] Uncertainty of measurement—Part 3: Guide to the expression of uncertainty in measurement (GUM). JCGM 100 [Internet]. 2008. Available from: https://www.bipm.org/utils/common/documents/jcgm/JCGM_100_2008_E.pdf [Accessed: 10-01-2019]

[12] ILAC Policy on Traceability of Measurement Results. ILAC P10:01/2013 [Internet]. 2013. Available from: https://ilac.org/publications-and-resources/ilac-policy-series/ [Accessed: 10-01-2019]

[13] ISO/IEC 17025:2017. General requirements for the competence of testing and calibration laboratories. Switzerland: ISO/IEC; 2017. p. 30

[14] ISO/IEC 17043:2010. Conformity Assessment. General requirements for proficiency testing. Switzerland: ISO/IEC; 2010. p. 39

[15] Velichko ON. Traceability of measurement results at different levels of metrological work. Measurement Techniques. 2009;**52**(11):1242-1248. DOI: 10.1007/s11018-010-9428-7

[16] Cox MG. The evaluation of key comparison data. Metrologia. 2002;**39**:589-595. DOI: 10.1088/0026-1394/39/6/10

[17] Mana G, Massa E, Predescu M. Model selection in the average of inconsistent data: An analysis of the measured Planck-constant values. Metrologia. 2012;**49**:492-500. DOI: 10.1088/0026-1394/49/4/492

[18] Velychko O, Akhmadov O. Final report on COOMET key comparison of capacitance at 10 pF (COOMET.EM-K4). Metrologia. 2017;**54**(1A):01005. DOI: 10.1088/0026-1394/54/1A/01005

[19] Velychko O, Darmenko Y. Final report on COOMET key comparison of AC/DC voltage transfer reference (COOMET.EM-K6.a). Metrologia. 2016; **53**(1A):01011. DOI: 10.1088/0026-1394/53/1A/01011

[20] Velychko O, Shevkun S. Support of metrological traceability of capacitance measurements in Ukraine. Eastern-European Journal of Enterprise Technologies. 2017;**3**(9 (87)):4-10. DOI: 10.15587/1729-4061.2017.101897

[21] Velychko O, Shevkun S. A support of metrological traceability of inductance measurements in Ukraine. Eastern-European Journal of Enterprise Technologies. 2017;**5**(9 (89)):12-18. DOI: 10.15587/1729-4061.2017.109750

[22] Evaluation of the Uncertainty of Measurement in Calibration. EA-04/02 M [Internet]. 2013. Available from: https://european-accreditation.org/wp-content/uploads/2018/10/ea-4-02-m-rev01-september-2013.pdf [Accessed: 10-01-2019]

[23] Delahaye F, Witt TJ. Linking the results of key comparisons CCEM-K4 with the 10 pF results of EUROMET.EM-K4. Metrologia. 2002; **39**:01005

[24] Velychko O. Proposals for linking the results of key comparisons CCEM-K4 and COOMET.EM-K4. In: Proceedings of the Conference on Precision Electromagnetic Measurements (CPEM 2010); 2010; Daejeon; South Korea: CPEM; 2010; 5545263. pp. 414-415. DOI: 10.1109/CPEM.2010.5545263

[25] Dierikx E, Nestor A, Melcher J, Kölling A, Callegaro L. Final report on the supplementary comparison EURAMET.EM-S26: inductance measurements of 100 mH at 1 kHz (EURAMET project 816). Metrologia. 2012;**49**:01002

[26] Velychko O, Karpenko S. Linking results of key and supplementary comparisons of regional metrology organization for electrical power. International Journal of Metrology and Quality Engineering. 2016;**7**(3). DOI: 10.1051/ijmqe/2016014

[27] Velychko O. Linking results of key and supplementary comparisons of AC/DC voltage transfer standard. International Journal of Metrology and Quality Engineering. 2018;**9**:4. DOI: 10.1051/ijmqe/2018002

[28] Chunovkina A, Zviagin N, Burmistrova N. Interlaboratory comparisons. Practical approach for data evaluation. In: Proceedings of the XX IMEKO World Congress "Metrology for Green Growth"; 2012; Busan, Republic of Korea. IMEKO; 2012. p. 5

[29] Briggs P. Proficiency testing for calibration laboratories. In: Proceedings of the XX IMEKO World Congress "Metrology for Green Growth"; 2012; Busan, Republic of Korea. IMEKO; 2012. p. 5

[30] Beckert SF, Fischer GE. Interlaboratory comparison of roughness measurement: Application of Algorithm A of ISO 13528:2015 in determining the designated value and the standard deviation. XXII World Congress of the International Measurement Confederation (IMEKO 2018). Journal of Physics: Conference

Series. 2018;**1065**:082007:4. DOI: 10.1088/1742-6596/1065/8/082007

[31] Claudio J, Costa M. Brazilian energy interlaboratory program applicative. In: Proceedings of the XX IMEKO World Congress "Metrology for Green Growth"; 2012; Busan, Republic of Korea. IMEKO; 2012. p. 6

[32] Sandu I, Dragomir L. Interlaboratory comparison. In: Proceedings of the 15th IMEKO TC 4 Symposium on Novelties in Electrical Measurements and Instrumentations; 2007; Iasi, Romania. IMEKO; 2007. p. 4

[33] Sousa JJL, Leitão LTS, Costa MM, Faria MC. Considerations on the influence of travelling standards instability in an interlaboratory comparison program. In: Proceedings of the XX IMEKO World Congress "Metrology for Green Growth"; 2012; Busan, Republic of Korea. IMEKO; 2012. p. 4

[34] ISO 13528:2015. Statistical methods for use in proficiency testing by interlaboratory comparisons. Switzerland: ISO; 2015. p. 89

[35] Velychko O, Shevkun S, Gordiyenko T, Mescheriak O. Interlaboratory comparisons of the calibration results of time meters. Eastern-European Journal of Enterprise Technologies. 2018;**1/9**(91):4-11. DOI: 10.15587/1729-4061.2018.121089

[36] Velychko O, Gordiyenko T. Features of the processing of results and estimation of measurement uncertainty of inter-laboratory comparison for calibration laboratories. Information Processing Systems. 2018;**4**(155):77-83. DOI: 10.30748/soi.2018.155.10

[37] Velychko O, Isaiev V. Interlaboratory comparison in context of inappropriate results of voltage thermal converter calibration. Journal of Electrical Engineering and Information Technologies. 2018;**3**(1–2):5-12

[38] Velychko O, Gordiyenko T. Linking Results of International Comparisons of the National Standard and the National Inter-Laboratory Comparisons. XXII World Congress of the International Measurement Confederation (IMEKO 2018). Journal of Physics: Conference Series. 2018;**1065**(4):072004. DOI: 10.1088/1742-6596/1065/7/072004

Chapter 2

Self-Calibration of Precision $XY\theta_z$ Metrology Stages

Chuxiong Hu, Yu Zhu and Luzheng Liu

Abstract

This chapter studies the on-axis calibration for precision $XY\theta_z$ metrology stages and presents a holistic $XY\theta_z$ self-calibration approach. The proposed approach uses an artifact plate, specially designed with *XY* grid mark lines and angular mark lines, as a tool to be measured by the $XY\theta_z$ metrology stages. In detail, the artifact plate is placed on the uncalibrated $XY\theta_z$ metrology stages in four measurement postures or views. Then, the measurement error can be modeled as the construction of $XY\theta_z$ systematic measurement error (i.e. stage error), artifact error, misalignment error, and random measurement noise. With a new property proposed, redundance of the $XY\theta_z$ stage error is obtained, while the misalignment errors of all measurement views are determined by rigid mathematical processing. Resultantly, a least square-based $XY\theta_z$ self-calibration law is synthesized for final determination of the stage error. Computer simulation is conducted, and the calculation results validate that the proposed scheme can accurately realize the stage error even under the existence of various random measurement noise. Finally, the designed artifact plate is developed and illustrated for explanation of a standard $XY\theta_z$ self-calibration procedure to meet practical industrial requirements.

Keywords: $XY\theta_z$ stage, self-calibration, measurement system, least square, stage error

1. Introduction

Precision $XY\theta_z$ motion stages are ubiquitously utilized in industrial mechanical systems to meet the requirement of high-performance manufacture [1]. As automatical servo systems, these stages have both precision linear encoders and angle encoders for measurement and motion feedback control [2–7]. In practice, the measurement accuracy inevitably suffers from surface non-flatness and un-roundness, axis nonorthogonality, scale graduation nonuniformity, encoder installation eccentricity, read-head misalignment, and so on, which resultantly generate systematic measurement error, i.e. stage error. The stage error can in principle be eliminated through calibration technology [8–10]. Due to the difficulty on finding a more accurate standard tool in traditional calibration technologies, self-calibration technology has been developed with utilization of an artifact with mark positions not precisely known. As an alternative of intelligent calibration processes, self-calibration is an effective and economical approach especially for micro-/nano-level mechanical systems [11–14].

Existing self-calibration technologies were developed for *X*, *XY*, *XYZ*, and angular metrology stages, respectively. For example, Takac studied one-dimensional self-calibration and developed a scheme that made a set of tool graduation marks appear to have identical spacing with relative scale [15]. In [16], self-calibration method for single-axis dual-drive nanometer positioning stage was presented. In [17], an *XY* self-calibration strategy was presented for two-dimensional metrology stages, which used an artifact plate as assistance measured by three views to construct equations of stage error, misalignment error, and artifact error. Fourier transformation was employed in the scheme to meet the challenge of random measurement noise. This method is popularly followed by many engineers and researchers [18–21]. In [18], a self-calibration algorithm was developed to test the out-of-plane error of two-dimensional profiling stages. The algorithm suppresses artifact-related errors in consideration of the geometrical congruence of three profile measurement views. Computer simulation and experimental results both showed that the calibration accuracy was free from artifact imperfection and only minimally affected by random measurement errors. In [19], a self-calibration method was proposed for mapping the errors in XY plane and the squareness error between Z-axis and XY plane of the scanning probe microscopes. In [22, 23], self-calibration approach for three-dimensional metrology stages was completely provided with experimental validation.

On the other hand, lots of self-calibration technologies have been developed for angular metrology systems [14, 24] in US National Institute of Standards and Technology (NIST), National Metrology Institute of Japan (NMIJ), Germany's National Metrology Institute the Physikalisch-Technische Bundesanstalt (PTB), Korea Research Institute of Standards and Science, etc. Specifically, circle closure principle was frequently used to cross-calibrate index tables in NIST [25, 26]. A high-precision rotary encoder self-calibration system was built based on equal-division-averaged method and had been adopted as the angular national standard system in NMIJ [27, 28]. The equal-division-averaged method was also expanded for self-calibration of the scale error in an angle comparator [29]. In addition, a known prime factor algorithm-based method was presented for self-calibration of divided circles in PTB [30, 31].

In summary of previous self-calibration strategies, a systematic self-calibration strategy for calibration of $XY\theta_z$ metrology stage is seldom published up to present. To address this problem, we have proposed a preliminary framework to self-calibrate the $\mathrm{XY}\theta_z$ stage error in [32], assuming that the angular coordinate and the XY coordinate are uncorrelated while the XY stage error and θ_z stage error are solved separately. This assumption leads to the final $\mathrm{XY}\theta_z$ calibration being not in a uniform coordinate, which means that it is not a complete and accurate $\mathrm{XY}\theta_z$ self-calibration strategy. In this chapter, we further study the self-calibration of precision $XY\theta_z$ metrology stages and present a complete and accurate on-axis self-calibration approach. Specifically, a new artifact plate is designed as the assistant tool, and four measurement views of the designed artifact plate on the uncalibrated $XY\theta_z$ metrology stage are constructed to provide measurement information. The detailed specification of the artifact plate on the $XY\theta_z$ stage is shown in **Figure 1**. Combining with symmetry, transitivity, and circle closure principle, certain redundance of the $XY\theta_z$ stage error is established, while the misalignment errors of all measurement views are determined by rigid mathematical manipulation. Resultantly, a least square-based $XY\theta_z$ self-calibration law is proposed for the final determination of the stage error. Computer simulation is conducted, and the calculation results validate that scheme proposed in this paper can figure out the stage error rather accurately in the absence of random measurement noise. The self-calibration accuracy of the proposed scheme is also tested to meet the challenge of various random measurement noises, and the calibration results validate that the scheme can effectively alleviate the effects of random measurement noise. Finally, the

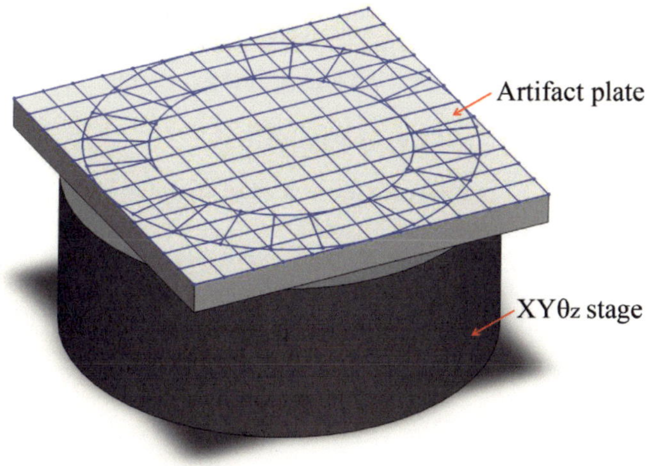

Figure 1.
An artifact plate with mark lines on an $XY\theta_z$ metrology stage.

designed artifact plate is manufactured, and a standard on-axis $XY\theta_z$ self-calibration procedure following the proposed scheme is introduced.

The proposed scheme mainly features the following two benefits: (1) Departing from previous self-calibration technologies, the proposed scheme first solves the on-axis self-calibration problem of $XY\theta_z$ metrology stages and (2) complicated mathematical manipulations, especially the calculations of misalignment errors in previous XY self-calibration schemes, are significantly avoided in the proposed strategy. The remainder of this chapter is organized as follows. In Section II, the stage error of $XY\theta_z$ metrology stage is explained, and a newly designed artifact plate and related artifact error are also described. The principle of the developed $XY\theta_z$ self-calibration scheme with four measurement views is presented in Section III. In Section IV, computer simulation is conducted to show the calibration performance of the proposed method. And the procedure for performing a standard $XY\theta_z$ self-calibration is presented in Section V. Finally, the conclusion is provided in Section VI.

2. Self-calibration problem formulation

2.1 Stage error

For a $XY\theta_z$ metrology system, linear encoders are employed for measuring movement along X and Y axes, and a rotary encoder for measuring rotation along θ_z axis. Thus, the systematic errors along X-axis, Y-axis, and θ_z axis are independent. And once the metrology system is set, the geometric relationship among X-axis, Y-axis, and θ_z axis is also determined, which will be described in detail later. In the Cartesian grid, define $\mathbf{G}_l(x,y)$ as the linear stage error at (x,y) where (x,y) is the true location. And $\mathbf{G}_r(\theta_z)$ is the rotary stage error at θ_z where θ_z is the true angle value. Herein, the uncalibrated $XY\theta_z$ field consists of XY with $L \times L$ and θ_z with 360°, while the $XY\theta_z$ origin point is set as the same at the center of the $L \times L$ field. In the following, we define

$$\begin{aligned}\mathbf{G}_l(x,y) &\equiv G_x(x,y)\mathbf{e}_x + G_y(x,y)\mathbf{e}_y \\ \mathbf{G}_r(\theta_z) &\equiv G_{\theta_z}(\theta_z)\mathbf{e}_{\theta_z}\end{aligned} \tag{1}$$

where $\mathbf{e}_x$, $\mathbf{e}_y$, and $\mathbf{e}_{\theta_z}$ are the unit vectors of the stage axes. For notation, we combine linear and rotary stage errors and define $\mathbf{G}(x, y, \theta_z)$ is the stage error at (x, y, θ_z) where (x, y) is the true location and θ_z is the true angle in the Cartesian grid:

$$\begin{aligned}\mathbf{G}(x,y,\theta_z) &\equiv \mathbf{G}_l(x,y) + \mathbf{G}_r(\theta_z) \\ &= G_x(x,y)\mathbf{e}_x + G_y(x,y)\mathbf{e}_y + G_{\theta_z}(\theta_z)\mathbf{e}_{\theta_z}\end{aligned} \tag{2}$$

Suppose the X-Y sample sites are in an $N \times N$ square array (N is odd) covering the $L \times L$ field and the θ_z sample lines are in a K array (K is a multiple of 4) covering the 360° field. Then in the Cartesian grid, the positions of the sample sites are

$$x_m = m\Delta, y_n = n\Delta, \theta_k = k\ominus \tag{3}$$

where $m = -\frac{N-1}{2}, -\frac{N-3}{2}, \ldots, \frac{N-1}{2}, n = -\frac{N-1}{2}, -\frac{N-3}{2}, \ldots, \frac{N-1}{2}$, and $\Delta = L/N$ which is called sample site interval; $k = 0, 1, 2, \ldots, K-1$ and $\ominus = 360/K°$. For notation simplicity, through Eq. (2), we can denote

$$\mathbf{G}_{m,n,k} \equiv G_{x,m,n}\mathbf{e}_x + G_{y,m,n}\mathbf{e}_y + G_{\theta_z,k}\mathbf{e}_{\theta_z} \tag{4}$$

where $G_{\theta_z,k} \equiv G_{\theta_z}(\theta_k)$, $G_{x,m,n} \equiv G_x(x_m, y_n)$, and $G_{y,m,n} \equiv G_y(x_m, y_n)$.

Similar to the detailed explanation in [17], to define the coordinates' origin, orientation, and grid scale of the XY axes stage, there are no translation property, no rotation property, and no magnification property for $G_{x,m,n}$ and $G_{y,m,n}$, which can be expressed mathematically as

$$\begin{aligned}&\sum_{m,n} G_{x,m,n} = \sum_{m,n} G_{y,m,n} = 0 \\ &\sum_{m,n}\left(G_{y,m,n}x_m - G_{x,m,n}y_n\right) = 0 \\ &\sum_{m,n}\left(G_{x,m,n}x_m + G_{y,m,n}y_n\right) = 0\end{aligned} \tag{5}$$

For X-Y axes, two dimensionless parameters O and R are defined as the XY nonorthogonality and the XY scale difference of $G_{x,m,n}$ and $G_{y,m,n}$, respectively. As a result, $G_{x,m,n}$ and $G_{y,m,n}$ are

$$\begin{aligned}G_{x,m,n} &= Oy_n + Rx_m + F_{x,m,n} \\ G_{y,m,n} &= Ox_m - Ry_n + F_{y,m,n}\end{aligned} \tag{6}$$

Therefore, one can obtain $G_{x,m,n}$ and $G_{y,m,n}$ by the first calculation of the first-order components O and R, and the later determination of the residual error $F_{x,m,n}$ and $F_{y,m,n}$. Noting that the origin of XY axes is the center of the sample array, we also can get the following properties of $F_{x,m,n}$ and $F_{y,m,n}$ which is also detailed in [17, 20]:

$$\begin{aligned}\sum_{m,n} F_{x,m,n} = \sum_{m,n} F_{x,m,n}x_m = \sum_{m,n} F_{x,m,n}y_n = 0 \\ \sum_{m,n} F_{y,m,n} = \sum_{m,n} F_{y,m,n}x_m = \sum_{m,n} F_{y,m,n}y_n = 0\end{aligned} \tag{7}$$

Besides, for rotary self-calibration, an important property, i.e. the circle closure principle, could directly bridge the gap between G_{k+K} and G_k, i.e. $G_{k+K} = G_k$ for $k = 0, 1, 2, \ldots, K-1$, which significantly facilitates the self-calibration process. To calculate the stage error components at θ_z, i.e. $G_{\theta_z,k}$, a new property must be pointed out as follows:

$G_{\theta_z,k}$ is definitely related to the errors of XY orientations. In other words, the expected value of angle deviation of points is exactly the θ_z orientation error of the radius vector where the points lie in. Angle deviation of points here means the angle between position vector of actual point and that of ideal point. Take $G_{\theta_z,0}$ as an example. For the point $(m, 0)$ on +X-axis, $\left(m = 1, 2, \cdots, \frac{N-1}{2}\right)$, define ϕ_m as the angle deviation of the point $(m, 0)$, i.e.

$$\begin{aligned}\phi_m &= <\left(x_m + G_{x,m,0}, y_0 + G_{y,m,0}\right), \left(x_m, y_0\right)> \\ &= <\left(x_m + G_{x,m,0}, G_{y,m,0}\right), \left(x_m, 0\right)>\end{aligned}$$

where $<\mathbf{a}, \mathbf{b}>$ is the angle between vectors **a** and **b** and, then,

$$G_{\theta_z,0} = E(\phi_m)$$

where $E(\phi_m)$ is the expectation of ϕ_m.

Noting that $G_{y,m,0} \ll x_m$ and $G_{x,m,0} \ll x_m$, one can obtain

$$\begin{aligned}\phi_m &= <\left(x_m + G_{x,m,0}, G_{y,m,0}\right), \left(x_m, 0\right)> \\ &= \arctan\left(\frac{G_{y,m,0}}{x_m + G_{x,m,0}}\right) \\ &\doteq \arctan\left(\frac{G_{y,m,0}}{x_m}\right) \\ &\doteq \frac{G_{y,m,0}}{x_m}\end{aligned}$$

which subsequently results in

$$G_{\theta_z,0} = E\left(\frac{G_{y,m,0}}{x_m}\right) \tag{8}$$

Similarly, along X-axis and Y-axis, we can obtain the following four equations:

$$\begin{aligned}G_{\theta_z,0} &= E\left(\frac{G_{y,m,0}}{x_m}\right) & \left(m = 1, 2, \cdots, \frac{N-1}{2}\right) \\ G_{\theta_z,\frac{K}{4}} &= E\left(-\frac{G_{x,0,n}}{y_n}\right) & \left(n = 1, 2, \cdots, \frac{N-1}{2}\right) \\ G_{\theta_z,\frac{K}{2}} &= E\left(\frac{G_{y,m,0}}{x_m}\right) & \left(m = -1, -2, \cdots, -\frac{N-1}{2}\right) \\ G_{\theta_z,\frac{3K}{4}} &= E\left(-\frac{G_{x,0,n}}{y_n}\right) & \left(n = -1, -2, \cdots, -\frac{N-1}{2}\right)\end{aligned} \tag{9}$$

The goal of the proposed self-calibration method is to determine $\mathbf{G}_{m,n,k}$ through different measurement postures, through which the measurement accuracy can be compensated directly.

2.2 Artifact error

In this chapter, an artifact plate which possesses mark lines different from previous researches in [17, 20] is designed specifically for $XY\theta_z$ self-calibration. **Figure 2** shows the details of artifact plate on the stage. In detail, an $N \times N$ grid mark array is on the artifact plate with the same size as the stage sample site array. Furthermore, it has K mark lines with equal angle interval. The plate $XY\theta_z$ coordinate axis' origin is located on the center of the mark array. During the plate movement, the plate axis will move with the plate. The locations of the nominal mark in the plate coordinate system are totally the same with that of the sample site in the stage coordinate system. Due to the unavoidable imperfection of the artifact plate, all the actual marks at (m, n, k) deviate from their nominal location by $\mathbf{A}_{m,n,k}$ which is defined as artifact error expressed by

$$\begin{aligned} \mathbf{A}_{m,n,k} &\equiv A_{x,m,n}\mathbf{e}_{px} + A_{y,m,n}\mathbf{e}_{py} + A_{\theta_z,k}\mathbf{e}_{pz} \\ A_{x,m,n} &\equiv A_x(x_m, y_n), A_{y,m,n} \equiv A_y(x_m, y_n), \\ A_{\theta_z,k} &\equiv A_{\theta_z}(\theta_k) \end{aligned} \tag{10}$$

where $m = -\frac{N-1}{2}, -\frac{N-3}{2}, \ldots, \frac{N-1}{2}, n = -\frac{N-1}{2}, -\frac{N-3}{2}, \ldots, \frac{N-1}{2}$, and $k = 0, 1, \ldots, K-1$; $\mathbf{e}_{px}$, $\mathbf{e}_{py}$, and $\mathbf{e}_{pz}$ are the unit vectors of the plate axes.

It should be noted that every mark on the artifact plate has an identification number (m, n, k). During the motions of the plate on the stage, the identification number of the mark will not change. This characteristic is also utilized to identify each physical mark of the plate in the following comparison of different measurement views. $A_{x,m,n}$ and $A_{y,m,n}$ also have no translation property and no rotation property [17, 20], which essentially have defined the axis origin and axis orientation, i.e.

$$\begin{aligned} \textstyle\sum_{m,n} A_{x,m,n} = \sum_{m,n} A_{y,m,n} = 0 \\ \textstyle\sum_{m,n}\left(A_{y,m,n}x_m - A_{x,m,n}y_n\right) = 0 \end{aligned} \tag{11}$$

3. $XY\theta_z$ self-calibration principle

3.1 The measurement views

The self-calibration method is based on four different postures or views of the designed artifact plate on the uncalibrated $XY\theta_z$ metrology stage, which is shown in **Figure 3**. As shown in **Figure 3**, the $XY\theta_z$ stage is the gray part, while the artifact plate is the white part. The 3-D specification can also be found in **Figures 1** and **2**.

Without the loss of generality, for each view, there inevitably exists a misalignment error; for that these coordinate axes cannot be aligned completely,

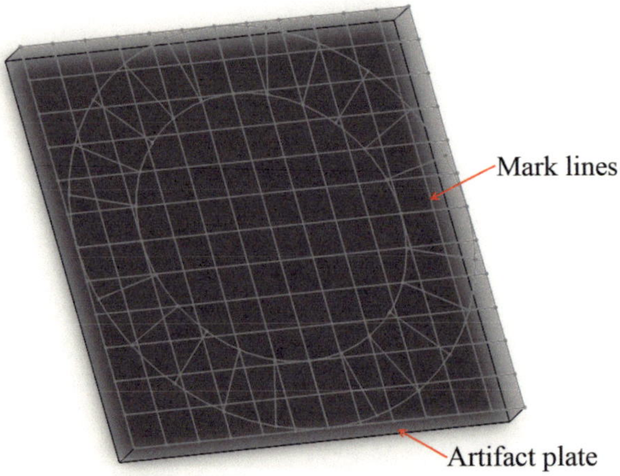

Figure 2.
A designed artifact plate with mark lines for $XY\theta_z$ self-calibration.

which will be considered as a misalignment error, which consists of a small rotation between their orientations and a small offset between their origins. Besides, random measurement noise also exists in the measurement process, but the effects of noise can be assumed to be completely attenuated by repeated measurements:

a. In View 0, which is the initial view, the $XY\theta_z$ axes of the plate are aligned as closely in line with those of the stage as possible; in this view, both grid and angular marks are measured.

b. In View 1, the artifact plate is rotated 90°, around the origin from View 0 on the stage; in this view, both grid and angular marks are measured.

c. In View 2, the artifact plate is rotated $360/K$°, i.e., ⊖, around the origin from View 0 on the stage; in this view, only angular marks are measured.

d. In View 3, the artifact plate is translated by one sample site, i.e., Δ, along +X-axis from View 0 on the stage; in this view, both grid and angular marks are measured.

For each measurement view, the artifact plate is firmly fixed on the stage, and a mark alignment system is needed to help the $XY\theta_z$ metrology stage to precisely measure the mark lines. The detailed instruments are presented later.

In the following, we would present the measurement and mathematical manipulations of each measurement view and then the reconstruction of the stage error map, in which V stands for the measured deviation for a mark from its nominal position in the stage coordinate.

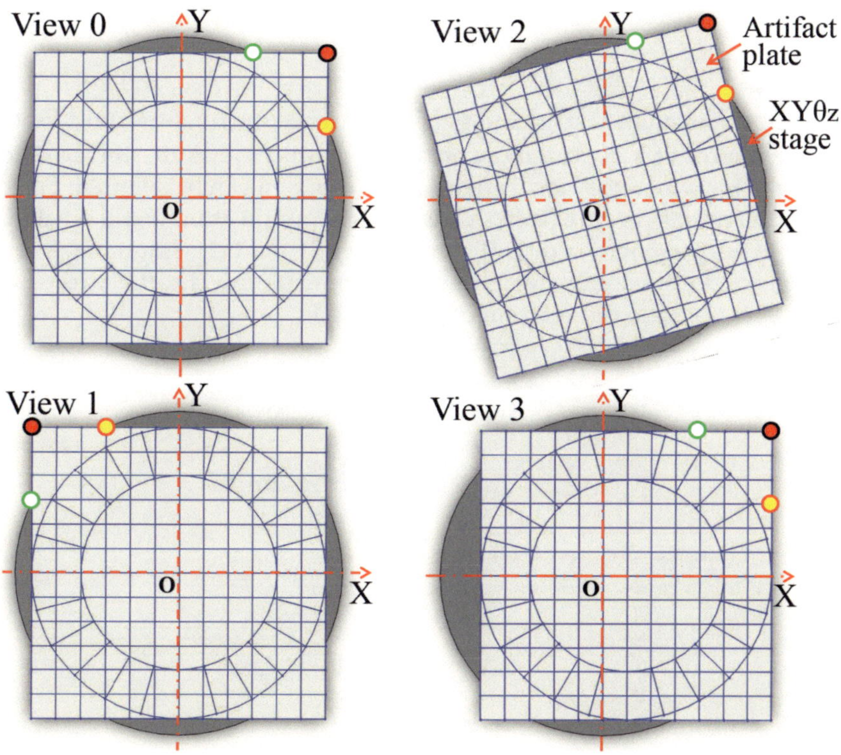

Figure 3.
Independent measurement views for XYθ$_z$ self-calibration.

For View 0,

$$\begin{aligned} V_{0,x,m,n} &= G_{x,m,n} + A_{x,m,n} - \varphi_0 y_n + t_{0x} \\ V_{0,y,m,n} &= G_{y,m,n} + A_{y,m,n} + \varphi_0 x_m + t_{0y} \\ V_{0,\theta_z,k} &= G_{\theta_z,k} + A_{\theta_z,k} + \varphi_0 \end{aligned} \tag{12}$$

where $m = -\frac{N-1}{2}, -\frac{N-3}{2}, \ldots, \frac{N-1}{2}$, $n = -\frac{N-1}{2}, -\frac{N-3}{2}, \ldots, \frac{N-1}{2}$, and $k = 1, 2, \ldots, K$.

For View 1,

$$\begin{aligned} V_{1,x,m,n} &= G_{x,-n,m} - A_{y,m,n} - \varphi_1 x_m + t_{1x} \\ V_{1,y,m,n} &= G_{y,-n,m} + A_{x,m,n} - \varphi_1 y_n + t_{1y} \\ V_{1,\theta_z,k} &= G_{\theta_z,k+\frac{K}{4}} + A_{\theta_z,k} + \varphi_1 \end{aligned} \tag{13}$$

where $m = -\frac{N-1}{2}, -\frac{N-3}{2}, \ldots, \frac{N-1}{2}$, $n = -\frac{N-1}{2}, -\frac{N-3}{2}, \ldots, \frac{N-1}{2}$, and $k = 1, 2, \ldots, K$.

For View 2,

$$V_{2,\theta_z,k} = G_{\theta_z,k+1} + A_{\theta_z,k} + \varphi_2 \tag{14}$$

where $k = 1, 2, \ldots, K$.

For View 3,

$$\begin{aligned} V_{3,x,m,n} &= G_{x,m+1,n} + A_{x,m,n} - \varphi_3 y_n + t_{3x} \\ V_{3,y,m,n} &= G_{y,m+1,n} + A_{y,m,n} + \varphi_3 x_m + t_{3y} \\ V_{3,\theta_z,k} &= G_{\theta_z,k} + A_{\theta_z,k} + \varphi_3 \end{aligned} \tag{15}$$

where $m = -\frac{N-1}{2}, -\frac{N-3}{2}, \ldots, \frac{N-3}{2}$, $n = -\frac{N-1}{2}, -\frac{N-3}{2}, \ldots, \frac{N-1}{2}$, and $k = 1, 2, \ldots, K$.

It shall be pointed out that φ_0 and $\mathbf{t}_0 = [t_{0x}, t_{0y}]$ are the rotation and offset of the misalignment error of View 0 and the notations of other views are similar. And φ_0 is a small angle, for which the 'small angle' approximation can be adopted. For the rotation misalignment error of other views, this approximation is still tenable.

Similar to the presentation of [17], combining (5) and (11), summing over all the sites of (12) and (13), we can obtain the misalignment error, i.e. the offset components $t_{0x}, t_{0y}, t_{1x}, t_{1y}$, and the rotation components φ_0, φ_1, i.e.

$$\begin{aligned} t_{0x} &= \frac{\sum_{m,n} V_{0,x,m,n}}{N^2}, \quad t_{0y} = \frac{\sum_{m,n} V_{0,y,m,n}}{N^2} \\ t_{1x} &= \frac{\sum_{m,n} V_{1,x,m,n}}{N^2}, \quad t_{1y} = \frac{\sum_{m,n} V_{1,y,m,n}}{N^2} \\ \varphi_0 &= \frac{\sum_{m,n} \left(V_{0,y,m,n} x_m - V_{0,x,m,n} y_n \right)}{\sum_{m,n} \left(x_m^2 + y_n^2 \right)} \\ \varphi_1 &= \frac{\sum_{m,n} \left(-V_{1,y,m,n} y_n - V_{1,x,m,n} x_m \right)}{\sum_{m,n} \left(x_m^2 + y_n^2 \right)} \end{aligned} \tag{16}$$

Noting Eqs. (5), (11), and (12), summing over all sites of (14) and (15), φ_2 and φ_3 can be determined, and the detailed result is

$$\varphi_2 = \frac{\sum_k V_{2,\theta_z,k} - \sum_k V_{0,\theta_z,k}}{K} + \frac{\sum_{m,n}\left(V_{0,y,m,n}x_m - V_{0,x,m,n}y_n\right)}{\sum_{m,n}\left(x_m^2 + y_n^2\right)}$$
$$\varphi_3 = \frac{\sum_k V_{3,\theta_z,k} - \sum_n V_{0,\theta_z,k}}{K} + \frac{\sum_{m,n}\left(V_{0,y,m,n}x_m - V_{0,x,m,n}y_n\right)}{\sum_{m,n}\left(x_m^2 + y_n^2\right)} \tag{17}$$

After elimination of the misalignment error in View 0, View 1, and View 2, combining Eq. (6), the measurement data are to be rearranged as:

$$U_{0,x,m,n} = V_{0,x,m,n} - t_{0x} + \varphi_0 y_n = F_{x,m,n} + A_{x,m,n} + Oy_n + Rx_m$$
$$U_{0,y,m,n} = V_{0,y,m,n} - t_{0y} - \varphi_0 x_m = F_{y,m,n} + A_{y,m,n} + Ox_m - Ry_n \tag{18}$$
$$U_{0,\theta_z,k} = V_{0,\theta_z,k} - \varphi_0 = G_{\theta_z,k} + A_{\theta_z,k}$$

$$U_{1,x,m,n} = V_{1,x,m,n} + \varphi_1 x_m - t_{1x} = F_{x,-n,m} - A_{y,m,n} + Ox_m - Ry_n$$
$$U_{1,y,m,n} = V_{1,y,m,n} + \varphi_1 y_n - t_{1y} = F_{y,-n,m} + A_{x,m,n} - Oy_n - Rx_m \tag{19}$$
$$U_{1,\theta_z,k} = V_{1,\theta_z,k} - \varphi_1 = G_{\theta_z,k+\frac{K}{4}} + A_{\theta_z,k}$$

$$U_{2,\theta_z,k} = V_{2,\theta_z,k} - \varphi_2 = G_{\theta_z,k+1} + A_{\theta_z,k} \tag{20}$$

And to keep the notation consistent with the previous views, we define

$$U_{3,x,m,n} = V_{3,x,m,n} + \varphi_3 y_n = F_{x,m+1,n} + A_{x,m,n} + Oy_n + Rx_m + \xi_x$$
$$U_{3,y,m,n} = V_{3,y,m,n} - \varphi_3 x_m = F_{y,m+1,n} + A_{y,m,n} + Ox_m - Ry_n + \xi_y \tag{21}$$
$$U_{3,\theta_z,k} = V_{3,y,m,n} - \varphi_3$$

where $\xi_x = t_{3x} + R\Delta$ and $\xi_y = t_{3y} + O\Delta$.

Comparing Eq. (18) of View 0 with Eq. (19) of View 1, with the same procedure in [17], the stage error components O and R can be calculated out as:

$$O = \frac{1}{2}\left[\frac{\sum_{m,n}\left(U_{0,x,m,n}y_n + U_{0,y,m,n}x_m\right)}{\sum_{m,n}\left(x_m^2 + y_n^2\right)} + \frac{\sum_{m,n}\left(U_{1,x,m,n}x_m - U_{1,y,m,n}y_n\right)}{\sum_{m,n}\left(x_m^2 + y_n^2\right)}\right]$$
$$R = \frac{1}{2}\left[\frac{\sum_{m,n}\left(U_{0,x,m,n}x_m - U_{0,y,m,n}y_n\right)}{\sum_{m,n}\left(x_m^2 + y_n^2\right)} + \frac{\sum_{m,n}\left(-U_{1,x,m,n}y_n - U_{1,y,m,n}x_m\right)}{\sum_{m,n}\left(x_m^2 + y_n^2\right)}\right] \tag{22}$$

3.2 $XY\theta_z$ self-calibration algorithm

A least square-based self-calibration algorithm is synthesized to determinate $G_{m,n,k}$. In this algorithm, the computation of the misalignment error components ξ_x and ξ_y is unnecessary, while φ_3 is determined by Eq. (17). Because O and R are known by Eq. (22), the algorithm is constructed to just calculate out $F_{x,m,n}$, $F_{y,m,n}$, and $G_{\theta_z,k}$.

Comparing Eq. (18) with (19), one obtains

$$F_{x,m,n} - F_{y,-n,m} = U_{0,x,m,n} - U_{1,y,m,n} - 2Oy_n - 2Rx_m$$
$$F_{y,m,n} + F_{x,-n,m} = U_{0,y,m,n} + U_{1,x,m,n} - 2Ox_m + 2Ry_n \tag{23}$$

Then combining Eqs. (18) and (21), define

$$\begin{aligned} L_{x,m,n} &= U_{3,x,m,n} - U_{0,x,m,n} = F_{x,m+1,n} - F_{x,m,n} + \xi_x \\ L_{y,m,n} &= U_{3,y,m,n} - U_{0,y,m,n} = F_{y,m+1,n} - F_{y,m,n} + \xi_y \end{aligned} \tag{24}$$

Consequently, we can obtain

$$\begin{aligned} & F_{x,m+2,n} - 2F_{x,m+1,n} + F_{x,m,n} = L_{x,m+1,n} - L_{x,m,n} \\ & F_{x,m+1,n+1} - F_{x,m+1,n} - F_{x,m,n+1} + F_{x,m,n} = L_{x,m,n+1} - L_{x,m,n} \\ & F_{y,m+2,n} - 2F_{y,m+1,n} + F_{y,m,n} = L_{y,m+1,n} - L_{y,m,n} \\ & F_{y,m+1,n+1} - F_{y,m+1,n} - F_{y,m,n+1} + F_{y,m,n} = L_{y,m,n+1} - L_{y,m,n} \end{aligned} \tag{25}$$

From previous subsections, Eqs. (7), (23), and (25) can yield

$$\begin{aligned} & \textstyle\sum_{m,n} F_{x,m,n} = \sum_{m,n} F_{x,m,n} x_m = \sum_{m,n} F_{x,m,n} y_n = 0 \\ & \textstyle\sum_{m,n} F_{y,m,n} = \sum_{m,n} F_{y,m,n} x_m = \sum_{m,n} F_{y,m,n} y_n = 0 \\ & F_{x,m,n} - F_{y,-n,m} = U_{0,x,m,n} - U_{1,y,m,n} - 2Oy_n - 2Rx_m \\ & F_{y,m,n} + F_{x,-n,m} = U_{0,y,m,n} + U_{1,x,m,n} - 2Ox_m + 2Ry_n \\ & F_{x,m+2,n} - 2F_{x,m+1,n} + F_{x,m,n} = L_{x,m+1,n} - L_{x,m,n} \\ & F_{x,m+1,n+1} - F_{x,m+1,n} - F_{x,m,n+1} + F_{x,m,n} = L_{x,m,n+1} - L_{x,m,n} \\ & F_{y,m+2,n} - 2F_{y,m+1,n} + F_{y,m,n} = L_{y,m+1,n} - L_{y,m,n} \\ & F_{y,m+1,n+1} - F_{y,m+1,n} - F_{y,m,n+1} + F_{y,m,n} = L_{y,m,n+1} - L_{y,m,n} \end{aligned} \tag{26}$$

which actually can determinate $F_{x,m,n}$ and $F_{y,m,n}$ with certain redundancy. Then a least square estimation law for $F_{x,m,n}$ and $F_{y,m,n}$ can be synthesized through a least square solution of the set of $F_{x,m,n}$ and $F_{y,m,n}$ equations to meet the challenge of random measurement noise [17, 20, 21]. Thus, combining the solved parameters O and R, we can obtain $G_{x,m,n}$ and $G_{y,m,n}$.

Afterward, according to (9), we can obtain the determination of $G_{\theta_z,0}$, $G_{\theta_z,\frac{K}{4}}$, $G_{\theta_z,\frac{K}{2}}$, and $G_{\theta_z,\frac{3K}{4}}$ with certain redundancy. Here we use the method of least squares as follows:

$$\begin{aligned} G_{\theta_z,1} &= \frac{\sum_{m=1}^{M} G_{y,m,0} x_m - \frac{1}{M}\sum_{m=1}^{M} G_{y,m,0}\sum_{m=1}^{M} x_m}{\sum_{m=1}^{M} x_m^2 - \frac{1}{M}\left(\sum_{m=1}^{M} x_m\right)^2} \\ G_{\theta_z,1+\frac{K}{4}} &= -\frac{\sum_{n=1}^{M} G_{x,0,n} y_n - \frac{1}{M}\sum_{n=1}^{M} G_{x,0,n}\sum_{n=1}^{M} y_n}{\sum_{n=1}^{M} y_n^2 - \frac{1}{M}\left(\sum_{n=1}^{M} y_n\right)^2} \\ G_{\theta_z,1+\frac{K}{2}} &= \frac{\sum_{m=-M}^{-1} G_{y,m,0} x_m - \frac{1}{M}\sum_{m=-M}^{-1} G_{y,m,0}\sum_{m=-M}^{-1} x_m}{\sum_{m=-M}^{-1} x_m^2 - \frac{1}{M}\left(\sum_{m=-M}^{-1} x_m\right)^2} \\ G_{\theta_z,1+\frac{3K}{4}} &= -\frac{\sum_{n=-M}^{-1} G_{x,0,n} y_n - \frac{1}{M}\sum_{n=-M}^{-1} G_{x,0,n}\sum_{n=-M}^{-1} y_n}{\sum_{n=-M}^{-1} y_n^2 - \frac{1}{M}\left(\sum_{n=-M}^{-1} y_n\right)^2} \end{aligned} \tag{27}$$

where $M = \frac{N-1}{2}$. Noting Eqs. (18), (19), and (20), we consequently obtain

$$\begin{aligned} G_{\theta_z, k+\frac{K}{4}} - G_{\theta_z, k} &= U_{1, \theta_z, k} - U_{0, \theta_z, k} \\ G_{\theta_z, k+1} - G_{\theta_z, k} &= U_{2, \theta_z, k} - U_{0, \theta_z, k} \end{aligned} \tag{28}$$

where $k = 1, 2, \ldots, K$ and $G_{\theta_z, k+K} = G_{\theta_z, k}$.

Combining (27) and (28), the stage error $G_{\theta_z, k}$ can be determined through a least square solution as the above equation group has K unknowns and 2K + 4 equations.

The proposed method features certain benefits remarked as follows:

- In previous self-calibration schemes for *XY* stages and *XYZ* stages [17, 18, 20–23], the properties of no translation and no rotation for the stage error cannot be used in View 3. Resultantly, it needs complicated algebraic manipulations to determine the misalignment error component φ_3. In this chapter, Eq. (17) directly determines the value of φ_3, and it is so convenient that complicated algebraic manipulations are significantly avoided.

- We propose a new property to construct connection between *XY* orientation and θ_z orientation. By a least square method, values of $G_{\theta_z, 1}$, $G_{\theta_z, \frac{K}{4}}$, $G_{\theta_z, \frac{K}{2}}$, and $G_{\theta_z, \frac{3K}{4}}$ are determined by Eq. (27), which is quite important for the calculation algorithm of $G_{\theta_z, k}$. With full utilization of the measurement of View 0, View 1, and View 2, we can construct a simple algorithm with strong robustness.

4. Computer simulation

In this section, we use MATLAB software to simulate the self-calibration process. First, arbitrary stage linear error maps on a 11×11 sample site array with the sample site interval $\Delta = 10$ mm are generated using the command 'normrnd' with a mean of 0 and standard deviation of 0.2 μm, which are utilized as the nominal $G_{x, m, n}$ and $G_{y, m, n}$. Besides, any stage rotary error is mapped within a revolution, while the sample site internal is $\Delta = 15°$. And the nominal $G_{\theta_z, k}$ is generated by mean value of 0 and standard deviation of 0.01°. And minor modification has been made to the data to satisfy the relevant requirements like Eqs. (5), (9), and (11). Then, arbitrary artifact linear error maps are generated with mean of 0 and standard deviation of 0.3 μm, which are employed as the nominal $\mathbf{A}_{m, n}$, and arbitrary artifact rotary error maps are generated with a mean of 0 and standard deviation of 0.01°, which are utilized as the nominal $A_{\theta_z, k}$. The nominal stage error component $\mathbf{G}_{m, n}$ is shown in **Figure 4** where the red lines are $\mathbf{G}_{m, n} \times 10000$. The nominal stage error component $G_{\theta_z, k}$ is shown in **Figure 5** where $G_{\theta_z, k}$ between the actual measurement system and perfect measurement system has been zoomed in for $360/\pi$ times. In addition, for each view, we add a random misalignment which is made up of a rotation and an offset. And the standard deviation value in the misalignment is 0.3° for rotation and 30 μm for offset. Since Eq. (26) has some redundance, it is clear that $\mathbf{G}_{m, n, k}$ can be figured out rather accurately if there is no random measurement noise. In addition, as there are no reported complete on-axis *XY*θ_z self-calibration strategies in published papers, we just test their own effectiveness of the proposed strategy. Herein, we focus on testing the calibration accuracy of the proposed strategy in various random measurement noises.

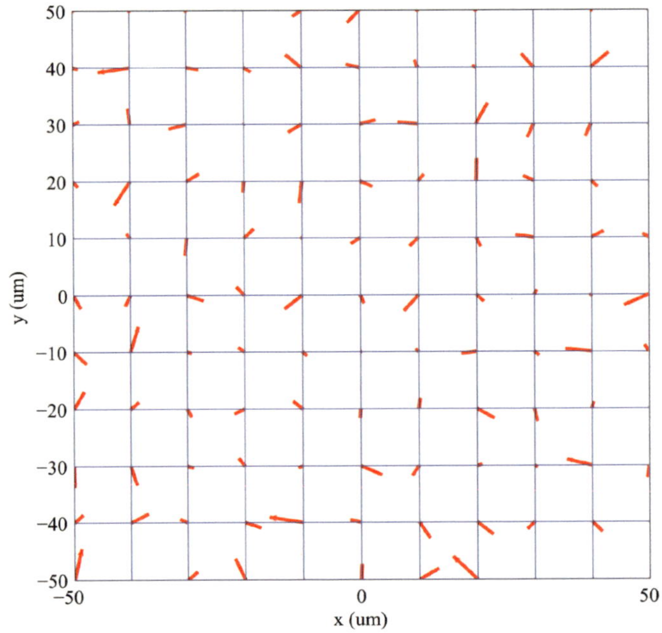

Figure 4.
$G_{m,n} \times 10000$ with $max(G_{m,n})$ = 0.5494 μm, $min(G_{m,n})$ = −0.5806 μm.

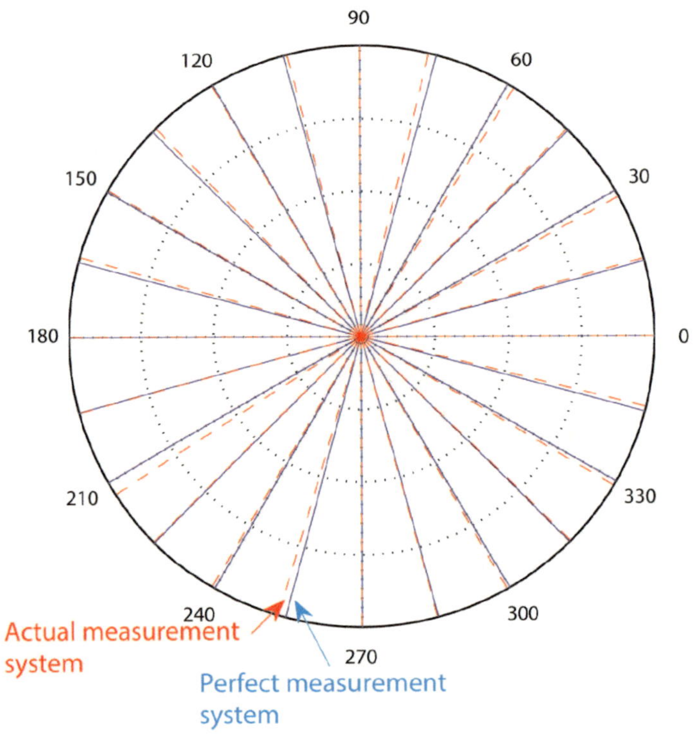

Figure 5.
$G_{\theta_z,k} \times 360/\pi^\circ$ with $max(G_{\theta_z,k})$ = 0.0243°, $min(G_{\theta_z,k})$ = −0.01282°.

4.1 Simulation with random measurement noise of standard deviation 0.02 μm for $G_{x,m,n}$ and $G_{y,m,n}$ and standard deviation 0.001° for $G_{\theta_z,k}$

For the simulation in this subsection, we add an independent random Gaussian measurement noise to every grid mark's each measurement and independent angular measurement noise to every angular mark's each measurement. $G_{x,m,n}$ and $G_{y,m,n}$'s random measurement noise is generated by a mean value 0 and standard deviation 0.02 μm, and $G_{\theta_z,k}$'s random measurement noise is generated by a mean value 0 and standard deviation 0.001°. The reconstructed stage error $\mathbf{G}_{m,n,k}$ can thus be determined according to the proposed four measurement views and self-calibration algorithm. So as to test the robustness of the algorithm, the calibration errors E_{Gx}, E_{Gy}, and $E_{G\theta}$ are calculated, which are defined as the deviation between the actual and the recalculated stage error, i.e. $E_{Gx} = G_{x,m,n} - \hat{G}_{x,m,n}$, $E_{Gy} = G_{y,m,n} - \hat{G}_{y,m,n}$, and $E_{G\theta} = G_{\theta,k} - \hat{G}_{\theta,k}$. In **Table 1**, the maximum, the minimum, and the standard deviation for E_{Gx}, E_{Gy}, and $E_{G\theta}$ are all listed. It is obvious that the stage error can be accurately recalculated by the proposed method even in the case of random measurement noise—when there is measurement noise by standard deviation 0.02 μm and 0.001° and the standard deviations of the calibration errors are smaller than 0.02 μm and 0.001°, respectively.

In addition, the algorithm's accuracy is also tested to meet the challenge of various random measurement noises. We generate the random measurement noises for 20 times. As a result, the calibration errors' standard deviations are shown in **Figures 6** and 7. All the results illustrate that all the 20 standard deviations keep in the same level with the measurement noises themselves. The simulation results demonstrate that the algorithm is robust and accurate. In addition, it can deal with the challenge of random measurement noise effectively.

4.2 Simulation with random measurement noise of standard deviation 0.002 μm for $G_{x,m,n}$ and $G_{y,m,n}$ and standard deviation 0.0001° for $G_{\theta_z,k}$

The simulation in this subsection is set up exactly the same way as in the previous subsection, except for adding a different random Gaussian measurement noise to every site's measurement, which is to test the consistency of the proposed scheme's robustness to random measurement noise. The random measurement noise for $G_{x,m,n}$ and $G_{y,m,n}$ is generated with a mean of 0 and standard deviation of 0.002 μm, and the random measurement noise for $G_{\theta_z,k}$ is generated with a mean of 0 and standard deviation of 0.0001°. Through the proposed scheme, the reconstructed stage error $\mathbf{G}_{m,n,k}$ can be calculated out, and the maximum value $max(\cdot)$, the minimum value $min(\cdot)$, and the standard deviation $std(\cdot)$ of E_{Gx}, E_{Gy}, and $E_{G\theta}$ are detailed in **Table 2**. Furthermore, the algorithm's accuracy and robustness are tested for arbitrary 20 times; the results are shown in **Figures 8** and **9**. It can be observed that the calibration error is also about the same size as the random measurement noises themselves. All these results further verify that the proposed

	$max(\cdot)$	$min(\cdot)$	$std(\cdot)$
E_{Gx} (μm)	0.0541	−0.0490	0.0195
E_{Gy} (μm)	0.0511	−0.0506	0.0196
$E_{G\theta}$ (°)	1.5877e−003	−1.2065e−003	7.1184e−004

Table 1.
Calculation performance indexes (with random measurement noise std = 0.02 μm).

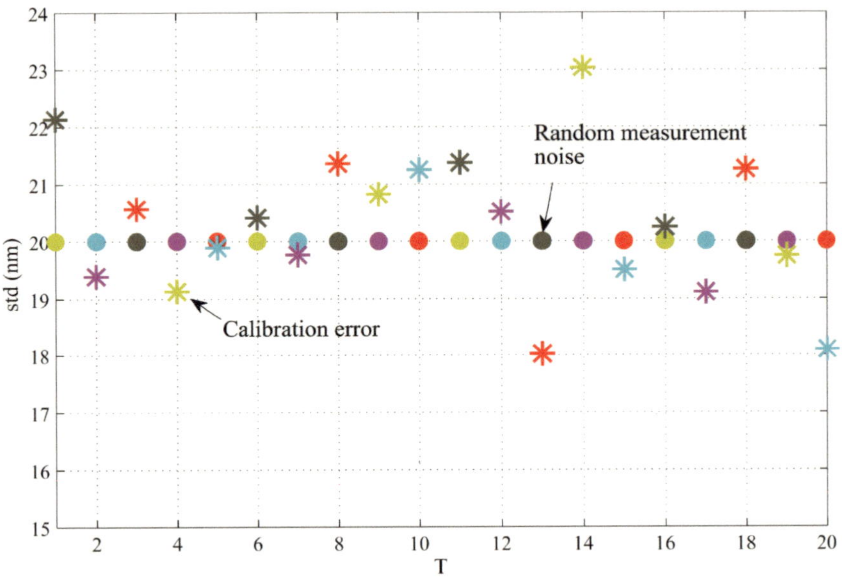

Figure 6.
Standard deviation of calibration error E_{Gx} & E_{Gy} for arbitrary 20 times (with random measurement noise std = 0.02 μm).

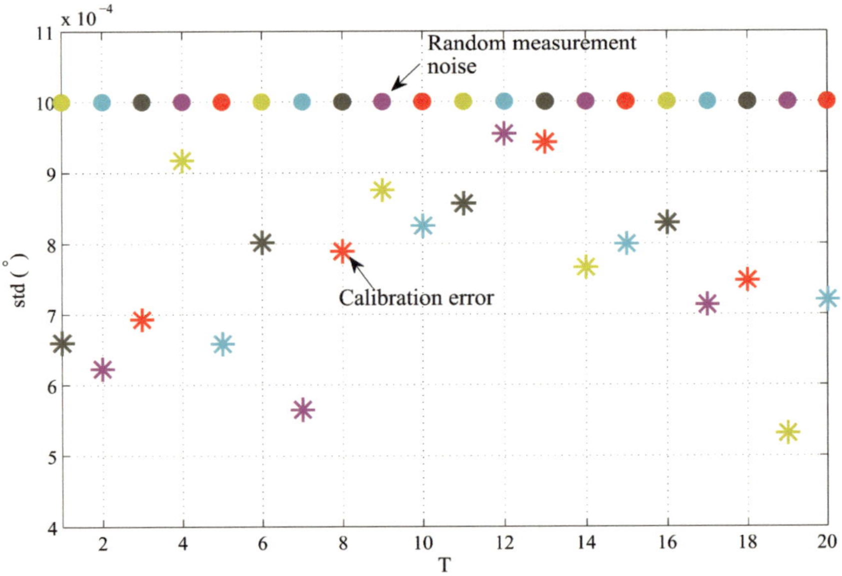

Figure 7.
Standard deviation of calibration error $E_{G\theta}$ for arbitrary 20 times (with random measurement noise std = 0.001°).

	max(·)	*min*(·)	*std*(·)
E_{Gx} (μm)	0.0051	−0.0051	0.0020
E_{Gy} (μm)	0.0051	−0.0053	0.0020
$E_{G\theta}$ (°)	2.9963e−004	−6.9733e−004	7.6657e−005

Table 2.
Calculation performance indexes (with random measurement noise std = 0.02 μm).

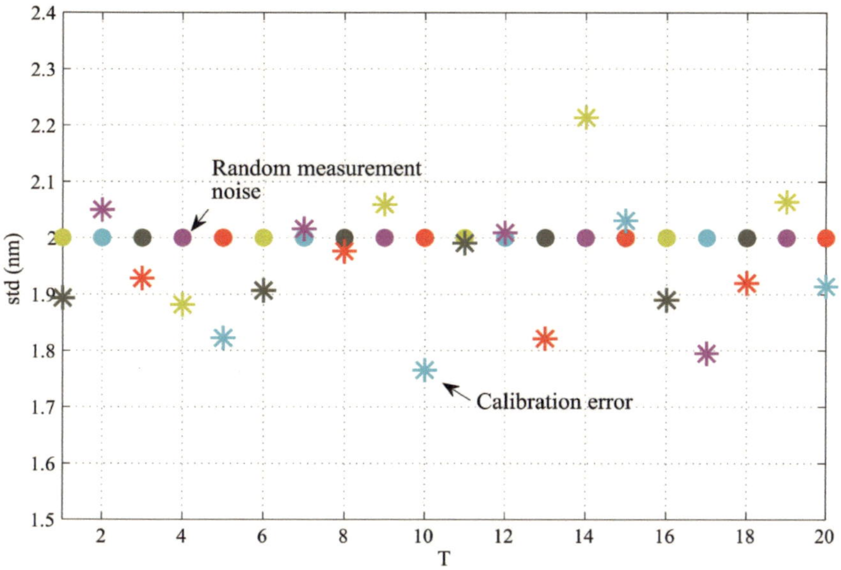

Figure 8.
Standard deviation of calibration error E_{Gx} & E_{Gy} for arbitrary 10 times (with random measurement noise std = 0.002 μm).

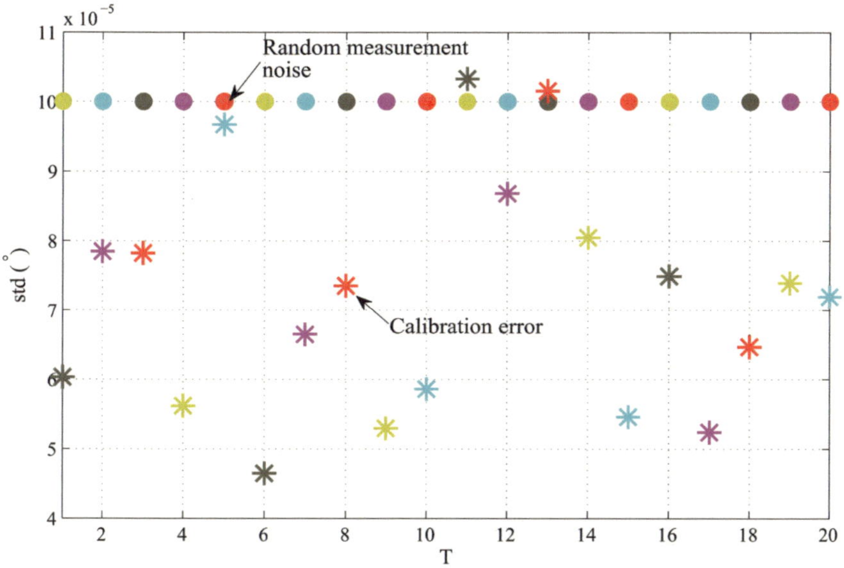

Figure 9.
Standard deviation of calibration error $E_{G\theta}$ for arbitrary 20 times (with random measurement noise std = 0.0001°).

strategy can accurately determine the stage error even under the existence of random measurement noise.

4.3 The artifact plate and the procedure for performing a standard $XY\theta_z$ self-calibration

Based on the proposed theory principle, we design and manufacture an artifact plate for $XY\theta_z$ self-calibration. The specification of the practical artifact plate is shown in **Figure 10**. Specifically, the artifact plate is made in square shape with

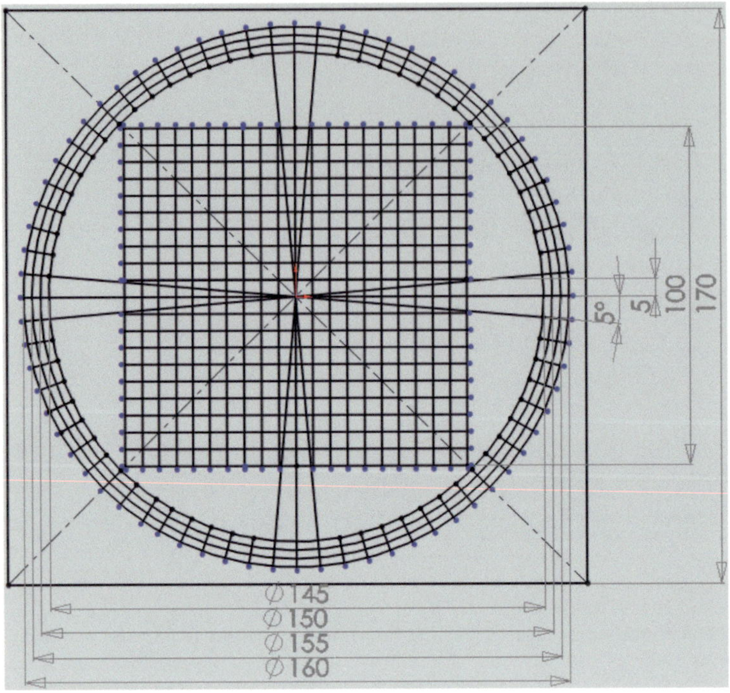

Figure 10.
A designed artifact plate for $XY\theta_z$ self-calibration.

material of optical glass. The thickness is 3 mm. The origin of the plate is located at the center. The radius or the circle in the plate is 80 mm. The circle of 360° is divided into 72 sectors by 72 lines with a width of 5 μm and accuracy of 5". The 21×21 sample site array is with the sample site interval $\Delta = 5$ mm. The accuracy of the circle lines and the straight lines is ± 1 μm and the width is 5 μm.

For presentation, convenience, and clarity, an example of an artifact plate on a $XY\theta_z$ stage is also provided and shown in **Figure 11**. In the following, we list the procedure of performing a standard $XY\theta_z$ self-calibration following the proposed scheme, which may be useful for engineers in practical implementations:

Step 0: As shown in **Figures 3** and **11**, put the artifact plate shown in **Figure 10** in the $XY\theta_z$ stage. The artifact plate's array marks are consequently at the stage's origin

Figure 11.
An artifact plate on a $XY\theta_z$ stage for self-calibration.

reference location, which is named as View 0 of **Figure 3**. The artifact plate's mark locations are measured in the $XY\theta_z$ metrology stage with the help of some mark alignment system such as those in [21, 23]. The mark alignment system should be designed to obtain the measurement data of each mark position accurately. Get $V_{0,x,m,n}$, $V_{0,y,m,n}$, and $V_{0,\theta_z,k}$ for P (e.g. $P = 5$) times, and average the results, and then obtain $U_{0,x,m,n}$, $U_{0,y,m,n}$, and $U_{0,\theta_z,k}$ by Eq. (18).

Step 1: After rotating 90° around the origin shown as View 1 in **Figure 3**, the artifact plate is replaced to the rotated reference location on the stage. The $XY\theta_z$ metrology stage is utilized to measure the artifact plate mark locations. Obtain $V_{1,x,m,n}$, $V_{1,y,m,n}$, and $V_{1,\theta_z,k}$ for P (e.g. $P = 5$) times and average the results. Get $U_{1,x,m,n}$, $U_{1,y,m,n}$, and $U_{1,\theta_z,k}$ by Eq. (19). And calculate out O and R by Eq. (22).

Step 2: After rotating $360/K°$ around the origin shown as View 2 in **Figure 3**, the artifact plate is replaced to the rotated reference location in the stage. The $XY\theta_z$ metrology stage is utilized to measure the angles of the artifact plate mark lines, and get $V_{2,\theta_z,k}$ for P (e.g. $P = 5$) times, and average the results, and then obtain $U_{2,\theta_z,k}$ by Eq. (20).

Step 3: After translating one grid interval Δ along +X-axis of the stage shown as View 3 in **Figure 3**, the artifact plate is replaced to the translated reference location in the stage. The $XY\theta_z$ metrology stage is used to measure the location of the marks of the artifact plate, and get $V_{3,x,m,n}$, $V_{3,y,m,n}$, and $V_{3,\theta_z,k}$ for P (e.g. $P = 5$) times, and average the results. Obtain $U_{3,x,m,n}$, $U_{3,y,m,n}$, and $U_{3,\theta_z,k}$ by Eq. (21).

Step 4: After above steps, the equation groups (26), (27), and (28) can be obtained. Therefore, a least square solution can be used for the determination of $F_{x,m,n}$, $F_{y,m,n}$, and $G_{\theta_z,k}$. Then $G_{x,m,n}$ and $G_{y,m,n}$ can be determined by Eq. (6) as O and R are previously computed in Step 1, which completes the final determination of $\mathbf{G}_{m,n,k}$.

5. Conclusions

In this chapter, an on-axis self-calibration approach has been first developed for precision $XY\theta_z$ metrology stages to solve the calibration problem. The proposed scheme uses a new artifact plate designed with special mark lines as the assistant tool for calibration. In detail, the artifact plate is placed on the uncalibrated $XY\theta_z$ metrology stages in four measurement postures or views. Then, the measurement error can be modeled as the construction of $XY\theta_z$ systematic measurement error (i.e. stage error) and other errors or noise. Based on the redundance of the $XY\theta_z$ stage error, a least square-based $XY\theta_z$ self-calibration law is resultantly synthesized for final determination of the stage error. Computer simulations have been conducted to verify that the proposed method can figure out the stage error rather accurately even under existence of various random measurement noises. Finally, a standard on-axis $XY\theta_z$ self-calibration procedure with the designed artifact plate is introduced. As an integration of XY self-calibration and θ self-calibration, the proposed scheme solves the $XY\theta_z$ self-calibration problem without complicated mathematical processing for misalignment errors. The developed approach has essentially provided a fundamental principle for on-axis self-calibration of precision $XY\theta_z$ metrology stages in practical applications.

The proposed scheme has pros and cons. It provides a significant theory fundamental for self-calibration of $XY\theta_z$ metrology or motion stages. However, the calibration accuracy is seriously affected by the mark alignment systems which may be a little complex. Therefore, the proposed self-calibration scheme is very suitable for standard calibration in national standard institutes but a little limited for wide industrial applications. In the next step, the development of this calibration system is an important topic, which will do help for wide applications of the proposed scheme.

Acknowledgements

This work was supported in part by the National Natural Science Foundation of China under Grant 51775305 and 51475262, Autonomous Scientific Research Project of Tsinghua University under Grant 20151080363, and Autonomous Research Project of State Key Lab of Tribology at Tsinghua University under Grant SKLT2018C02.

Author details

Chuxiong Hu[1,2*], Yu Zhu[1,2] and Luzheng Liu[1,2]

1 State Key Lab of Tribology, Department of Mechanical Engineering, Tsinghua University, Beijing, China

2 Beijing Key Lab of Precision/Ultra Precision Manufacture Equipments and Control, Tsinghua University, Beijing, China

*Address all correspondence to: cxhu@tsinghua.edu.cn

References

[1] Zhong G, Shao Z, Deng H, Ren J. Precise position synchronous control for multi-axis servo systems. IEEE Transactions on Industrial Electronics. 2017;**64**(5):3707-3717

[2] Hu C, Wang Z, Zhu Y, et al. Accurate three-dimensional contouring error estimation and compensation scheme with zero-phase filter. International Journal of Machine Tools and Manufacture. 2018;**128**:33-40

[3] Hu C, Hu Z, Zhu Y, Wang Z. Advanced GTCF based LARC contouring motion control on an industrial X-Y linear-motor-driven stage with experimental investigation. IEEE Transactions on Industrial Electronics. 2017;**64**(4):3308-3318

[4] Chen Z, Yao B, Wang Q. μ-Synthesis based adaptive robust control of linear motor driven stages with high-frequency dynamics: A case study with comparative experiments. IEEE/ASME Transactions on Mechatronics. 2015; **20**(3):1482-1490

[5] Chen Z, Yao B, Wang Q. Accurate motion control of linear motors with adaptive robust compensation of nonlinear electromagnetic field effect. IEEE/ASME Transactions on Mechatronics. 2013;**18**(3):1122-1129

[6] Hu C, Wang Z, Zhu Y, Zhang M, Liu H. Performance oriented precision LARC tracking motion control of a magnetically levitated planar motor with comparative experiments. IEEE Transactions on Industrial Electronics. 2016;**63**(9):5763-5773

[7] Liu G, Dass R, Nguang S-K, Partridge A. Principles, design, and calibration for a genre of irradiation angle sensors. IEEE Transactions on Industrial Electronics. 2013;**60**(1): 210-216

[8] Li X, Shamsi P. Model predictive current control of switched reluctance motors with inductance auto-calibration. IEEE Transactions on Industrial Electronics. 2016;**63**(6): 3934-3941

[9] Evans CJ, Hocken RJ, Estler WT. Self-calibration: Reversal, redundancy, error separation, and 'absolute testing. Annals of the CIRP. 1996;**35**(2):617-632

[10] Zhao Y, Lin Y, Xi F. Calibration-based iterative learning control for path tracking of industrial robots. IEEE Transactions on Industrial Electronics. 2015;**62**(5):2921-2929

[11] Takac MT, Ye J, Raugh MR, et al. Self-calibration in two dimensions: The experiment. In: Proceedings of SPIE Metrology, Inspection, and Process Control for Microlithog.; Bellingham. 1996. pp. 130-146

[12] Raugh MR. Two-dimensional stage self-calibration: Role of symmetry and invariant sets of points. Journal of Vacuum Science & Technology B. 1997; **15**(6):2139-2145

[13] De Cecco M, Debei S, Zaccariotto M, et al. A self-calibration method for nonorthogonal angles between gimbals of rotational inertial navigation system. IEEE Transactions on Industrial Electronics. Nov. 2015; **62**(4):2352-2362

[14] Lu X-D, Trumper DL. Self-calibration of on-axis rotary encoders. CIRP Annals: Manufacturing Technology. 2007;**56**(1):499-504

[15] Takac MT. Self-calibration in one dimension. In: Proceedings of SPIE 13th Annual BACUS Symposium on Photomask Technology and Management; Santa Clara. 1993. pp. 80-86

[16] Jeong YH, Dong J, Ferreira PM. Self-calibration of dual-actuated single-axis nanopositioners. Measurement Science and Technology. 2008;**19**(4):1-13

[17] Ye J, Takac MT, Berglund CN, et al. An exact algorithm for self-calibration of precision metrology stages. Precision Engineering. 1997;**20**(1):16-32

[18] Yoo S, Kim SW. Self-calibration algorithm for testing out-of-plane errors of two-dimensional profiling stages. International Journal of Machine Tools & Manufacture. 2004;**44**:767-774

[19] Xu M, Dziomba T, Dai G, Koenders L. Self-calibration of scanning probe microscope: Mapping the errors of the instrument. Measurement Science and Technology. 2008;**19**(2):1-6

[20] Hu C, Zhu Y, Hu J, Zhang M, Xu D. A holistic self-calibration algorithm for XY precision metrology systems. IEEE Transactions on Instrumentation and Measurement. 2012;**61**(9):2492-2500

[21] Zhu Y, Hu C, Hu J, Yang K. Accuracy and simplicity oriented self-calibration approach for two-dimensional precision stages. IEEE Transactions on Industrial Electronics. 2012;**60**(6):2264-2272

[22] Dang QC, Yoo S, Kim S-W. Complete 3-D self-calibration of coordinate measuring machines. Annals of the CIRP. 2006;**55**(1):1-4

[23] Hu C, Zhu Y, Hu J, Xu D, Zhang M. A holistic self-calibration approach for determination of three-dimensional stage error. IEEE Transactions on Instrumentation and Measurement. 2013;**62**(2):483-494

[24] Lu X-D, Graetz R, Amin-Shahidi D, Smeds K. On-axis self-calibration of angle encoders. CIRP Annals: Manufacturing Technology. 2010;**59**(1): 529-534

[25] Estler T, Queen H. An advanced angle metrology system. Annals of the CIRP. 1993;**42**(1):573-576

[26] Estler WT. Uncertainty analysis for angle calibrations using circle closure. Journal of Research of the National Institute of Standards and Technology. 1998;**103**(2):141-151

[27] Masuda T, Kajitani M. An automatic calibration system for angular encoders. Precision Engineering. 1989;**11**(2): 95-100

[28] Watanabe T, Fujimoto H, Masuda T. Self-calibratable rotary encoder. In: Proceedings of the 7th Int., Sym, Meas., Techol. Intellig. Instrum. 2005. pp. 240-245

[29] Kim J-A, Kim JW, Kang C-S, Jin J, Eom TB. Precision angle comparator using self-calibration of scale errors based on the equal-division-averaged method. In: MacroScale 2011 – Proceedings. Recent Developments in Traceable Dimensional Measurements. 2011. pp. 1-4

[30] Probst R. Self-calibratable of divided circles on the basis of a prime factor algorithm. Measurement Science and Technology. 2008;**19**(1):1-11

[31] Just A, Krause M, Probst R, Bosse H, Haunerdinger H, Spaeth C, et al. Comparison of angle standards with the aid of a high-resolution angle encoder. Precision Engineering. 2009;**33**:530-533

[32] Zhu Y, Hu C, Hu J, Zhang M, Xu D. On-axis self-calibration of precision XYθ_z metrology systems: An approach framework. In: Proceedings of 2013 IEEE/ASME International Conference on Advanced Intelligent Mechatronics (AIM). 2013. pp. 1078-1083

Chapter 3

Third-Order Nonlinear Optical Properties of Quantum Dots

Khalil Ebrahim Jasim

Abstract

Quantum dots (QDs) are semiconducting nanocrystalline particles. QDs are attractive photonic media. In this chapter, we introduce third-order nonlinear optical properties and a brief idea about the physics of QDs. Z-scan technique and theoretical analysis adopted to obtain nonlinear parameters will be discussed. Analysis of third-order nonlinear optical parameters for PbS QDs suspended in toluene with radii 2.4 and 5.0 nm under different excitation beam power level and three different wavelengths (488, 514, and 633 nm) will be detailed. Third-order optical susceptibility $\chi^{(3)}$ and optical-limiting behavior of PbS QD suspended in toluene are presented. Irrespective of their size, QDs are a good example of optical limiters with low threshold.

Keywords: quantum dot, semiconductor, nonlinear optics, Z-scan technique, nonlinear index of refraction, nonlinear absorption coefficient, third-order optical susceptibility, optical limiting, optical switching

1. Introduction

Generally speaking, quantum dots (QDs) are nanocrystalline structures that can be grown using physical or chemical methods. QDs are an essential medium for extensive range of applications in photonics. Some attention-grabbing applications of these materials include fluorescence imaging, electroluminescence, frequency up conversion lasing, stabilization of laser power fluctuations, photon microscopy, solar cells, optical signal reshaping, and nonlinear optics. QDs demonstrate a strong confinement for both electrons and holes due to size effect (quantum confinement). Therefore, nonlinear optical properties are expected to be greatly enhanced in QD media, due to flexibility of their electronic and optical properties that are controlled via choosing the precise size and concentration in the system. This chapter will discuss the physics and capacity of QDs as a nonlinear medium. Introductory idea about nonlinear optics will be outlined. Then, theoretical background of the Z-scan technique used to investigate QDs' third-order nonlinear optical properties will be detailed. As an example, characterization of lead sulfide (PbS)'s nonlinear optical properties will be presented.

2. Physics of quantum dots

In bulk structure of semiconductors, charge carriers (electrons and holes) have continuous distribution of energy states and their motion is not confined. However, in quantum well (QW) structures, these charge carries are confined and their energy states are quantized in one dimension. Further confinement and quantization of energy in two dimensions is achieved in quantum wire structures where motion of charge carriers is confined and energy states are quantized in two dimensions. Confinement of electron in three dimensions and hence quantization of its energy states can be established in quantum dot structures. QDs can be treated as zero-dimensional bulk solids. QDs are nanostructured semiconductor crystalline media. In QD structure, the density of states can be represented by a delta function, as illustrated in **Figure 1a**. In this system, only discrete energy levels are allowed with a specific wave vector **k** value for each allowed energy state (**Figure 1b** and **c**).

The energy gap of a QD can be estimated as the sum of the bandgap energy, the confinement energy, and the bound exciton energy. The bandgap energy is determined by the size of the quantum dot. In strong confinement regime, the size of the QD is smaller than the exciton Bohr radius a_{EBR}, where the energy levels split up. In this case, we have

$$a_{EBR} = \varepsilon_r \left(\frac{m}{\mu}\right) a_B \tag{1}$$

where m is the mass, μ is the reduced mass, ε_r is the size-dependent dielectric constant, and $a_B = 0.53 \times 10^{-10}$ m is the Bohr radius. **Table 1** shows values of a_{EBR} for some common semiconductors [1].

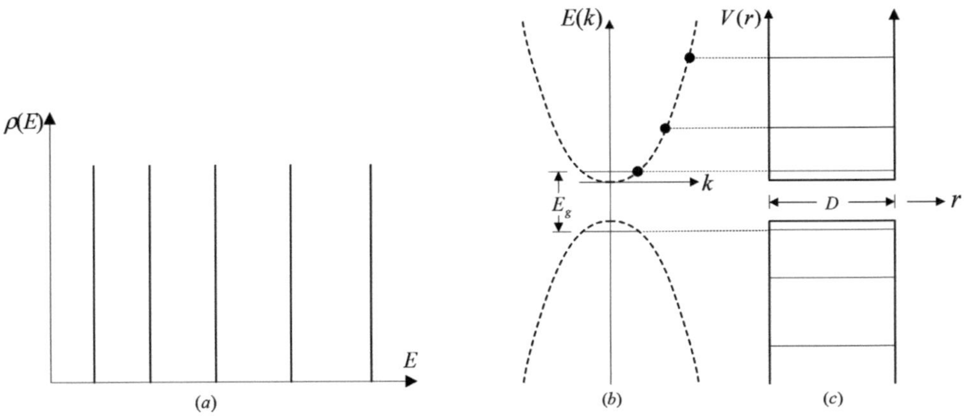

Figure 1.
(a) Energy density of states in QDs is represented by delta functions; (b) due to confinement effect, the energy gap is larger than that of the bulk material; (c) allowed energy in QDs is ideally that of an infinite quantum well.

Semiconductor structure	Exciton Bohr radius (nm)	Bandgap energy (eV)
PbS	40.0	0.41
GaAs	28.0	1.43
CdTe	15.0	1.50
CdSe	10.6	1.74
ZnSe	8.4	2.58
CdS	5.6	2.53

Table 1.
Exciton Bohr radius and bandgap energy of some common semiconductors [1].

The total energy released in the emission can thus be written as

$$E_{total} = E_{Bandgap} + E_{Confinment} + E_{Exciton} \tag{2}$$

where $E_{Bandgap}$ is the bulk bandgap. Hence, according to [2], the last two terms in Eq. (2) are,

$$E_{Confinment} = \frac{\eta^2 \pi^2}{2R^2}\left(\frac{1}{m_e} + \frac{1}{m_h}\right) = \frac{\eta^2 \pi^2}{2\mu R^2} \tag{3}$$

and

$$E_{exciton} = -\frac{1.8 e^2}{2\pi \varepsilon_0 \varepsilon_r R} \tag{4}$$

In the above equations, m_n and m_p are the electron and hole effective masses, respectively, $\varepsilon_0 = 8.8545 \times 10^{-12}$ F/m is the permittivity of free space, and R is the QD radius. Other quantities are as defined above.

The above simple model shows that the energy of QDs is dominated by the quantum confinement effect (Eq. 3), where the energy varies inversely with the square of the QD's radius. **Figure 2** shows absorbance spectrum obtained using dual-beam spectrophotometer for PbS QDs with different radii (sizes); as the size of the QD decreases, its bandgap energy increases (blue shift of the excitonic peak).

It is interesting to note that QDs unlike organic dyes when excited their fluorescence is much stronger and absorbance peak is widely spaced from that of fluorescence as presented in **Figure 3**, the peak emission wavelength is bell-shaped (Gaussian) and occurs at a slightly longer wavelength than the lowest energy exciton peak (the absorption onset). This energy separation is what is referred to as the Stokes shift. Therefore, there is less overlapping between absorbance and fluorescence spectra. Moreover, the QD's bleaching and damage threshold is much more than that of organic dyes.

Figure 4 shows fluorescence spectrum of excited PbS quantum dots suspended in toluene. As the excitation power increases, fluorescence intensity increases without any noticeable shift in the fluorescence peak.

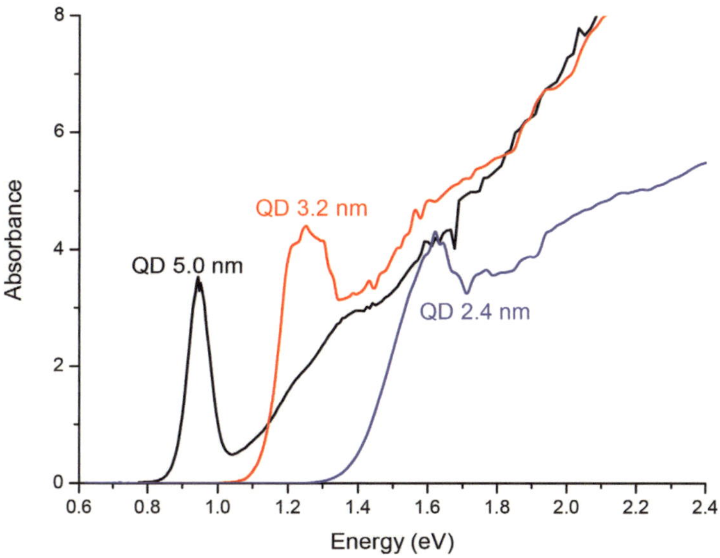

Figure 2.
Energy dependence versus absorbance of PbS QDs with three different sizes.

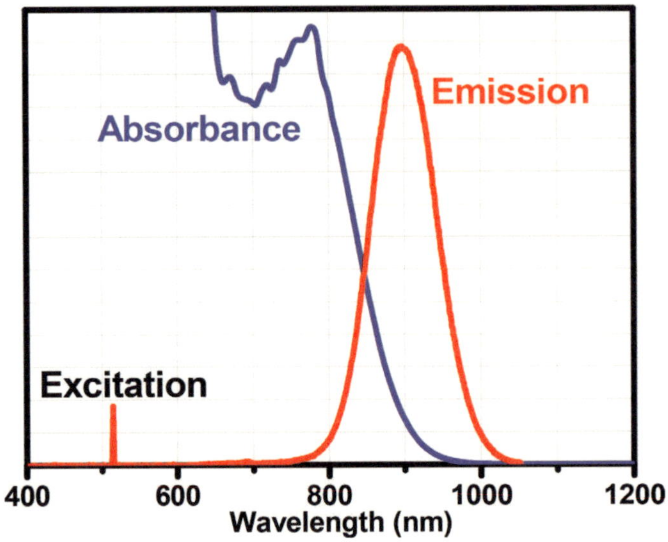

Figure 3.
Absorbance and fluorescence spectra for PbS QD of size 2.4 nm.

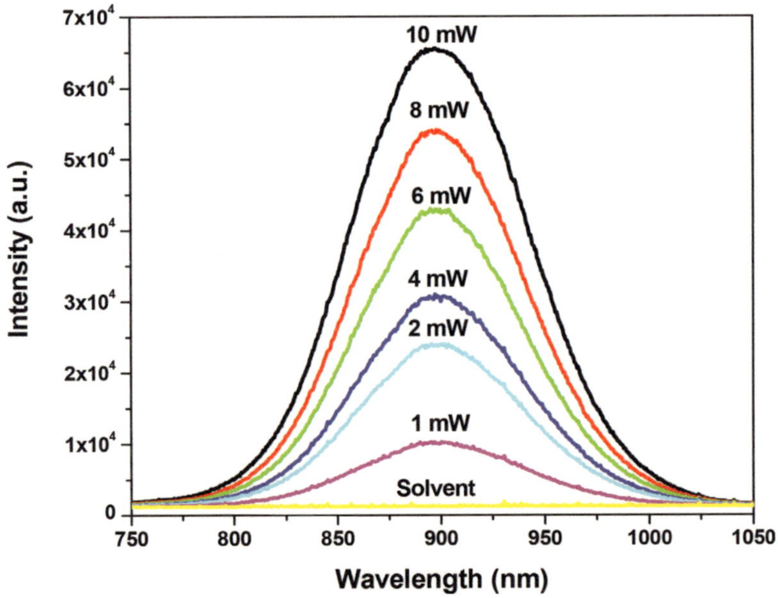

Figure 4.
Fluorescence spectra for PbS QDs of size 2.4 nm excited with Ar laser of λ = 514 nm at different beam power.

3. Nonlinear optics

Generally speaking, nonlinear optics is a branch in photonics concerned with studying optical properties and applications of nonlinear optical materials. Nonlinear optical media interact with electromagnetic waves in a manner different than what we get used to with ordinary materials such as glass slide or water. A linear optical material reflects, transmits, and absorbs part of a light beam incident on it without altering its nature such as color (frequency). Linear media respond to electromagnetic waves (EM) in a linear fashion. Therefore, it is a passive optical media. On the other hand, there are materials that upon interaction with EM waves of specific intensity become active and they behave as self-focusing (positive lens) or self-defocusing (negative lens), diffracting, beam fanning, or optical limiting media. Nonlinear optical media can be used to generate new frequencies.

3.1 Intensity-dependent refractive index and absorption coefficient

The result of the interaction between low-intensity non-coherent light beam and high-intensity coherent light beam differs as the normal (non-coherent) light interaction shows a linear response to the electromagnetic (EM) field and the optical characteristics of the material remain unchanged because they do not rely on the light intensity and the frequency remains constant. The occurrence of superposition is possible, but the waves do not interact with each other. On the other hand, the interaction between the laser (coherent) beam and the material shows a nonlinear response to the electromagnetic field and the optical characteristics of the material such as refractive index and the speed of light do change. Superposition principle does not apply because of interaction of light with light [3, 4]. Nonlinear refraction might occur from molecular reorientation, the electronic Kerr effect, or optically induced heating of the material [5–7].

The refractive index of NLO materials can be described by the following relation [8]:

$$n(I) = n_0 + n_2 I = n_0 + \Delta n \tag{5}$$

where n_0 is the linear refractive index, I is the incident optical field density or intensity (irradiance), and n_2 is a new optical constant which is called the second-order index of refraction (nonlinear refractive index coefficient). n_2 is the rate when the refractive index increases with increasing optical density. The change in refractive index Δn, in Eq. (5), which is proportional to the square of the applied electric field is called optical Kerr effect. The reported order of magnitude of n_2 for semiconductors is from 10^{-10} to 10^{-2} cm/W. It has been found that both magnitude and sign of n_2 are sensitive to both the wavelength and polarization state of the exciting laser beam [8]. The total refractive index n(I) is a linear function of intensity I.

The nonlinear absorption of NLO materials can be described by the following relation:

$$\alpha\,(I) = \alpha_0 + \beta\, I \tag{6}$$

where α_0 is the linear absorption coefficient, and β is the nonlinear absorption coefficient.

Both nonlinear index of refraction and nonlinear absorption coefficient are related to third-order optical susceptibility as given by the following relations, [9, 10]:

$$\mathrm{Re}\chi^{(3)}(esu) = \left(\frac{\varepsilon_0 c^2}{10^4 \pi}\right) n_0^2 n_2 \;\; (cm^2/W) \tag{7}$$

$$\mathrm{Im}\chi^{(3)}(esu) = \left(\frac{\varepsilon_0 c^2}{4 \times 10^2 \pi^2}\right) n_0^2 \lambda \beta \;\; (cm/W) \tag{8}$$

$$\left|\chi^{(3)}\right| = \sqrt{\left(\mathrm{Re}\chi^{(3)}\right)^2 + \left(\mathrm{Im}\chi^{(3)}\right)^2} \tag{9}$$

Here c is the speed of light in vacuum (3×10^8 m/s), ε_0 is the permittivity of free space (8.854×10^{-12} F/m), and n_0 is the linear index of the sample.

3.2 Third-order nonlinear processes

3.2.1 Self-phase modulation (SPM)

Self-phase modulation (SPM) is a nonlinear optical phenomenon of light-matter interaction which has been discovered in the early days of nonlinear optics. The strong field of a laser beam (coherent waves), when propagating in a third-order nonlinear medium, interacts with the medium and induces a varying refractive index of the medium due to the optical Kerr effect. This interaction imposes a phase shift in the laser pulse, leading to a change of the pulse's frequency spectrum [3, 8]. In brief, self-phase modulation is the intensity-dependent nonlinear refractive index that modulates the optical phase ($\Delta\phi$) of the laser beam electric field which is given by the following relation:

$$\Delta\phi = 2\pi n_2 L \frac{P}{A\lambda_o} \tag{10}$$

where P is the optical beam's power, L is the traveled distance in the medium, and A is the cross-sectional area. Laser beam has a finite cross section. However, due to SPM, laser beam will appear to have self-diffraction, which is responsible for the known nonlinear optical phenomena of self-focusing and self-defocusing. Moreover, the temporal variation of the laser intensity leads to SPM in time. As the time derivative of the phase of a wave is simply the angular frequency of the wave, SPM also appears as frequency modulation.

3.2.2 Self-focusing and self-defocusing effects

Another attention-grabbing phenomenon of light-matter interaction associated with SPM is self-focusing. It occurs whenever the strong field of a laser beam propagates in a third-order nonlinear medium whose refractive index depends on the beam intensity. The real part of the third-order susceptibility results in a change in the index of refraction of the material [7].

If the nonlinear refractive index is positive and the incident beam is of higher intensity in the center than the edge, then self-focusing will occur. In that case, the refractive index and the optical path length for rays at the center are greater than those at their edges. The same condition will occur for propagation via focusing lens and, because of that, the light beam will create its own positive lens in the nonlinear medium. Furthermore, when the beam focusing increases, the strength of the nonlinear lens increases resulting in stronger focusing and strength of the lens, this will result in catastrophic focusing where the beam is breaking down into a very intense small spot. This phenomenon may occur in materials that have transparency at high intensity such as crystals, glasses, gases, liquids, and plasmas [8].

In contrast, if the nonlinear refractive index is negative and the incident beam is of higher intensity in the center than the edge, then self-defocusing will occur. In that case, the refractive index and the short optical path length for rays at the center are smaller than those at their edges. The same condition will occur for propagation via negative focal length lens and beam defocuses [7].

3.2.3 Two-photon absorption (TPA)

Two-photon absorption is a nonlinear absorption process where two photons of identical or different frequencies are simultaneously absorbed when an intense

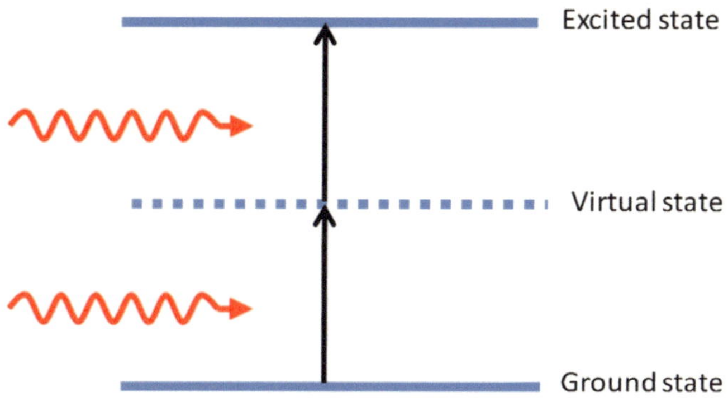

Figure 5.
Schematic illustration of two-photon absorption process.

coherent beam propagates in a third-order nonlinear medium as illustrated in **Figure 5** [11].

3.3 Optical limiting (OL)

Optical limiter is a device that exhibits a linear transmittance to power, irradiance, energy, or fluence below a threshold and clamps the output to a constant above it as shown in **Figure 6** [5, 11, 12], thereby providing safety to sensors or eyes because nonlinear absorption coefficient increases with increase in input power. Increase of sample temperature due to absorption results in nonlinear absorption coefficient and optical limiting effect [13]. To behave as an active optical limiter, the nonlinear optical medium must possesses high linear transmittance, low limiting threshold, fast response time, broadband response, and low optical scattering [14].

Researches on optical limiting (OL) effect started in the 1960s by Geusic et al and Ralston et al. with three-photon nonlinear absorption in inorganic semiconductors. Many demonstrations of optimized OL effect were accomplished through two-photon absorption (TPA) followed by subsequent excited state absorption (ESA).

Besides the effect of nonlinear refraction, nonlinear absorption is another kind of mechanisms used in OL. Reverse saturable absorption RSA and TPA are nonlinear absorption mechanisms used in OL [15, 16]. TPA-based devices are also proper for other applications such as optical power stabilization, optical pulse reshaping, and optical spatial field reshaping [5, 17]. Combining nonlinear

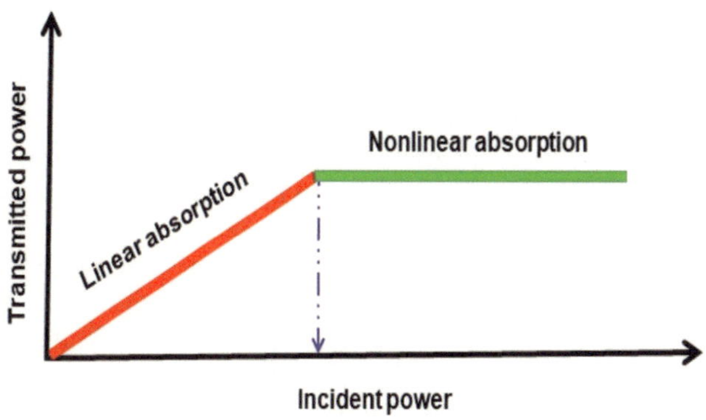

Figure 6.
Optical response of an optical limiter to incident power.

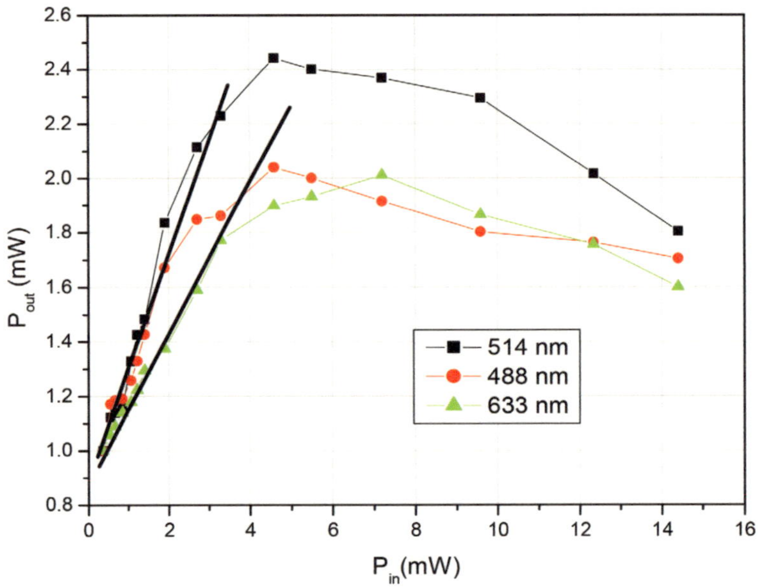

Figure 7.
Optical limiting response of PbS QDs of radius 5.0 nm excited at different wavelengths.

mechanisms has as well accomplished improvement in optical limiting, such as self-defocusing with TPA, TPA with excited state absorption (ESA). Inorganic materials combine ESA and RSA [12]. **Figure 7** shows optical limiting effect by PbS QDs of radius 5.0 nm excited with continuous wave (CW) Ar ion laser (λ = 488 and 514 nm) and HeNe laser (λ = 633 nm). It is obvious from **Figure 7** that OL threshold is wavelength dependent.

4. Characterization techniques

Many experimental techniques have been developed in order to achieve precise measurement of both nonlinear absorption coefficient (β) and nonlinear index of refraction coefficient (n_2) and then calculating the third-order susceptibility $\chi^{(3)}$. Examples of techniques that are in use are: nonlinear interferometry, degenerate four-wave mixing (DFWM), nearly degenerate three-wave mixing, ellipse rotation, and beam distortion measurements. However, a simple, and quite very accurate technique named Z-scan technique is widely in use and trusted (by majority of nonlinear optics researchers) to measure and tell the sign of both β and n_2, and hence calculating magnitude of $\chi^{(3)}$ [18–20]. Z-scan technique is a well-established technique in characterizing third-order nonlinearity mainly because it is experimentally easy to construct and optimized. The nature and sign of nonlinearity can be identified from the raw data. These data do not require elaborated mathematical analysis.

The main idea of the Z-scan technique is to measure the intensity change in incident Gaussian laser beam transmission intensity through the sample as a function of its position (Z-position) from the focus of the beam. Transmitted beam is collected by means of two detector arrangements: closed aperture (CA) in order to recognize the sign and measure n_2 as illustrated in **Figure 8**, and open aperture (OA) in order to measure β. The working principle of Z-scan technique is based on the motion of the sample along the so-called Z axis from –Z to +Z through the focus

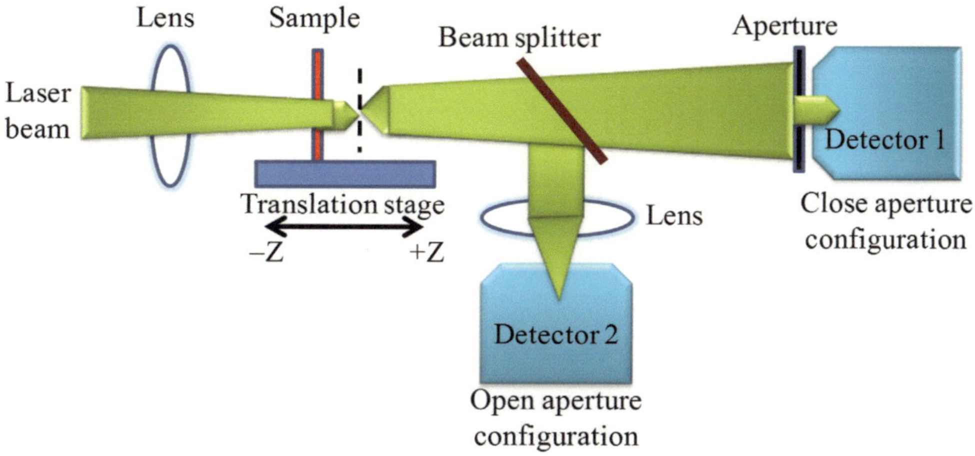

Figure 8.
Schematics of experimental setup used for the closed- and open-aperture Z-scan.

of a tightly focused Gaussian laser beam. The motion of the sample across the focused beam exposes it to a varying irradiance (intensity) level reaching its maximum value at the focus point as shown in **Figure 8**.

In this chapter, in order to apply theoretical models developed by Mansoor Sheik Bahae [18–20], samples must be prepared in the form of thin samples. A thin sample is an optical medium with thickness smaller than the diffraction length of a focused Gaussian beam.

4.1 Closed-aperture Z-scan configuration

As shown in **Figure 8**, if a thin sample is scanned with closed-aperture setup and Gaussian laser beam, both the sign and magnitude of nonlinear refractive index are determined. Considering a sample with negative nonlinear refractive index (self-defocusing), the scan starts far from focal length (–Z) and the sample exhibits negligible nonlinear refraction at low irradiance. As the sample gets closer to the focus beam, irradiance increases, and the sample starts acting as a negative lens. The transmittance through the aperture changes because of variation of sample location relative to the focus. When the sample becomes very close to the focus, it starts acting as a negative lens, and hence, the beam irradiance arriving at the aperture increases due to self-defocusing behavior. Conversely, as the sample moves away from the focus (+Z), the beam irradiance arriving at the detector decreases due to self-focusing effect. The transmittance becomes linear again when the scan ends. The result of this scan is represented in the transmittance versus position curve as shown in **Figure 9a**. A material possessing negative nonlinear refractive index (self-defocusing behavior) gives a transmittance with peak followed by a valley. However, sample with positive nonlinear index of refraction (self-focusing behavior) shows transmittance with valley followed by a peak as shown in **Figure 9b**. Thus, the sign of a material's nonlinearity can be determined directly from the transmittance versus position curve. This is a special feature of the Z-scan technique.

Although CA configuration is sensitive to nonlinear index of refraction, presence of nonlinear absorption can be detected on the transmittance curve. The presence of two-photon absorption (TPA) or reverse saturable absorption (RSA) enhances the valley and suppresses the peak as presented in **Figure 10a**. On the other hand, the presence of nonlinear absorption effect due to saturable absorption (SA) enhances the peak and suppresses the valley as shown in **Figure 10b** [18].

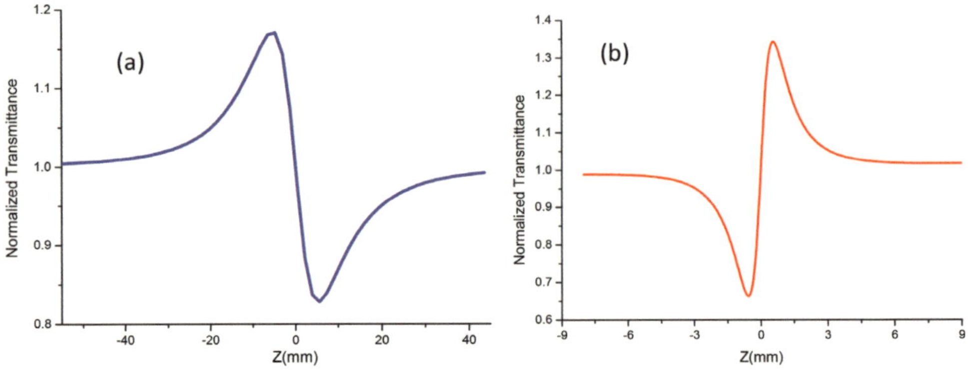

Figure 9.
Closed-aperture Z-scan normalized transmittance curve: (a) negative n_2, and (b) positive n_2.

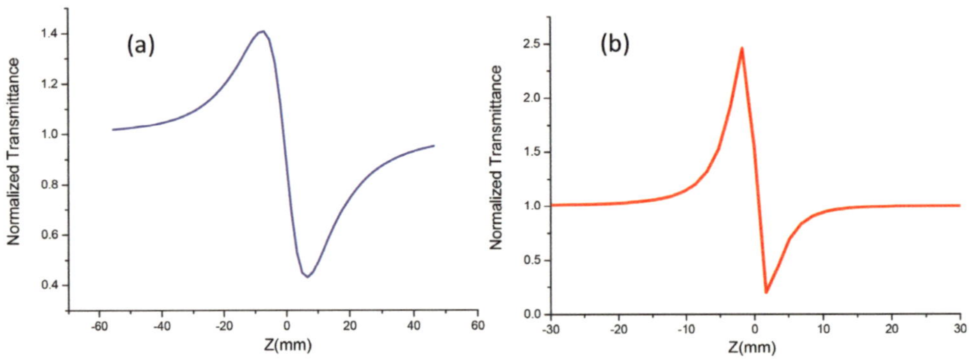

Figure 10.
Closed- aperture Z-scan effect of nonlinear absorption on transmittance of the sample in the presence of: (a) two photon absorption or reverse saturable absorption, and (b) saturable absorption.

4.2 Open-aperture Z-scan

When Z-scan is performed with open-aperture configuration by removing the aperture in front of the detector D2 shown in **Figure 8**, the nonlinear absorption coefficient is calculated from the normalized transmittance curve. Sample transmittance decreases (increases) as sample position Z gets closer to the focus point of the lens with minimum (maximum) value of the transmittance at the focal point, then it increases (decreases) as the sample moves away from the focus and reaches its linear behavior at far +Z position as illustrated in **Figure 11**.

In **Figure 11**, transmittance versus position curve has a symmetric shape around the focus; this is due to the symmetry of the Gaussian beam irradiance. If the transmittance is minimum at the focus (valley) as shown in **Figure 11**, this indicates the existence of two-photon absorption (TAP) or RSA with a positive type of nonlinear absorption. However, when the transmittance is maximum at the focus (peak), this indicates the presence of saturable absorption (SA) with negative type of nonlinear absorption [7, 11].

Z-scan technique requires a high-quality Gaussian beam for measurement. Tilting of sample during the scan affects the outcome of the scan. Also, the Z-scan technique is sensitive to all nonlinear optical mechanisms that give rise to a change of the refractive index and/or absorption coefficient. Thus, it is not possible to find out the original physical processes obtained from Z-scan technique.

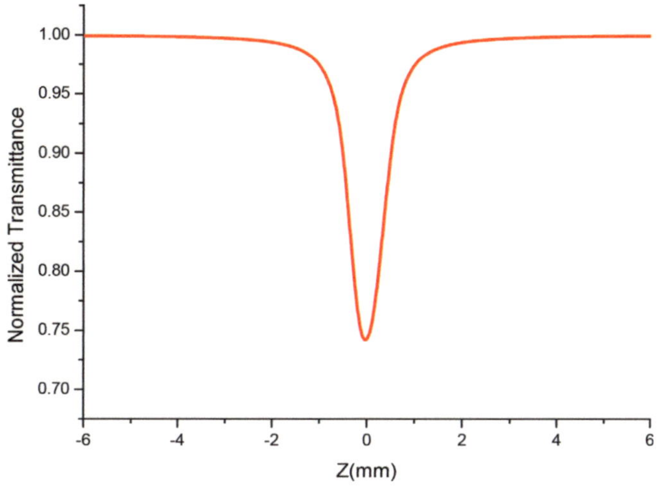

Figure 11.
Open-aperture Z-scan normalized transmittance curve.

4.3 Theoretical analysis

In nonlinear media, as a result of absorption of high-intensity excitation beam, the temperature in the surface of the sample exhibits a spatial distribution that creates a spatial variation of the refractive index that produces a thermal lens and hence a phase distortion of the propagating laser beam is produced. The normalized peak-to-valley transmittance ΔT_{pv} is linearly related to the phase distortion $\Delta\Phi_o$ through the following [21]:

$$\Delta T_{PV} = A\,\Delta\Phi_0 \tag{11}$$

where A is a constant given by:

$$A = 0.406\,(1\text{-S})^{0.25} \tag{12}$$

$$S = 1 - \exp\left(-2r_a^2/\omega_a^2\right) \tag{13}$$

where S is the aperture linear transmittance, r_a is the aperture radius, and ω_a is the beam spot size (radius) at the aperture. The phase distortion $\Delta\Phi_0$ is defined as

$$\Delta\Phi_0 = k n_2 I_o L_{eff} \tag{14}$$

$$I_o = 2P_o/\pi\omega_o^2 \tag{15}$$

where I_o is the intensity of the laser beam at the focus (z = 0), P_o is the instantaneous input laser power into the sample, k= $2\pi/\lambda$ is the wave number, n_2 is the nonlinear refractive index, and L_{eff} is the effective thickness of the sample,

$$L_{eff} = [1 - \exp(-\alpha_o L)]/\alpha_o \tag{16}$$

where α_0 is the linear absorption coefficient and L is the thickness of the sample. The nonlinear absorption coefficient β is given by [22]

$$\beta = \frac{2\sqrt{2}T_{\min}}{I_o L_{eff}} \tag{17}$$

where T_{min} is the normalized transmittance of the open aperture.

The third-order nonlinear optical susceptibility $\chi^{(3)}$ can be calculated once the nonlinear refractive index n_2 and the nonlinear absorption coefficient β are known using Eqs. 7–9 . The suitability of the material for all-optical switching can be evaluated using the figure of merit [23]:

$$W = n_2 I / \alpha_0 \lambda \tag{18}$$

The values W >> 1 are ideal for applications in all-optical switching.

In many situations, absorption and nonlinear index are mixed together (as shown in **Figure 10**). Consequently, one needs to adopt the Gaussian decomposition (GD) method, which was implemented by Sheik-Bahae and his co-workers [18–20]. This method consists of some complex mathematical steps to decompose the complex electric field emerging from the sample into a summation of Gaussian beams through Taylor series expansion of the nonlinear phase term. Since only small phase changes are considered, few terms of Taylor expansions are used. The normalized Z-scan transmittance T(z) can be calculated by using the following equation:

$$T(z) = \frac{\int_{-\infty}^{+\infty} P_T(z, \Delta\phi_0(t))dt}{S \int_{-\infty}^{+\infty} P_0(t)dt} \tag{19}$$

where P_T $(z, \Delta\phi_0(t))$ is the transmitted power through the aperture, $P_0(t)$ (Eq. 15) is the instantaneous input power (within the sample), and S is linear transmittance through the aperture. T(z) does not depend on the geometry or the wavelength while the size of aperture is an effective parameter. In case of large aperture (S = 1) the effect will disappear in T(z) in all z and $\Delta\phi_0$.

In some cases, material nonlinear absorption is weak and nonlinear refraction is more pronounced such that the transmittance curve obtained in closed-aperture configuration shows symmetric peak and valley. Therefore, one can fit the normalized transmittance curves to calculate the nonlinear index of refraction using the following equation:

$$T(z) = 1 + \frac{4x\Delta\phi_0}{(x^2+9)(x^2+1)} \tag{20}$$

where $x = z/z_R$, $z_R = \pi\omega_0^2/\lambda$ is the Rayleigh range or the diffraction length of the Gaussian beam, ω_0 is the beam's waist, and $\Delta\phi_0$ is the on-axis phase shift due to nonlinear refraction expressed as:

$$\Delta\phi_0 = kn_2 I_0 L_{eff} \tag{21}$$

Nonlinear absorption coefficient β can be calculated by fitting the open-aperture normalized transmittance curve using the following equation:

$$T(z) = 1 - \frac{\Delta\Psi_0}{(x^2+1)} \tag{22}$$

where $\Delta\Psi_0$ is the on-axis phase shift due to nonlinear absorption expressed as:

$$\Delta\Psi_0 = \frac{I_0\, L_{eff}}{2\sqrt{2}}\beta \tag{23}$$

In many situations, both nonlinear refraction and nonlinear absorption are mixed such that the closed-aperture transmittance shows asymmetric peak-valley curve like the one presented in **Figure 10**. In this case, the normalized transmittance curves are fitted using the following equation [15]:

$$T(z) = 1 + \frac{4x\Delta\phi_0}{(x^2+9)(x^2+1)} - \frac{2(x^2+3)\Delta\Psi_0}{(x^2+9)(x^2+1)} \tag{24}$$

The nonlinear refractive index and nonlinear absorption coefficient are calculated from Eqs. (21) and (23).

In cases of high single-photon absorption using CW laser even at few mill watts, thermal lensing becomes pronounced such that the Gaussian decomposition model will not be suitable to fit data since the closed-aperture normalized transmittance shows a very slow return to the normal transmittance as the sample is scanned further away from focal point (+Z) (for example see **Figure 12**). Therefore, a modified model has been suggested by [24], called the thermal-lens model. The normalized transmission of closed-aperture Z-scan is given by:

$$T(z) = \left(1 + \Delta\phi_0 \frac{2x}{(x^2+1)} + \Delta\phi_0^2 \frac{2x}{(x^2+1)}\right)^{-1} \tag{25}$$

Figure 12 shows the importance of adopting thermal model to get proper fitting of the collected data.

However, as shown in **Figure 13**, thermal lens model fails to fit properly the closed-aperture normalized transmittance data for the same PbS quantum dots (radius 2.4 nm) when excitation power P_0 = 6 mW but wavelength is 633 nm. Application of the Gaussian decomposition model resulted in better fitting of data. In conclusion, the investigator in order to accurately calculate third-order nonlinear parameters must chose the most proper fitting model.

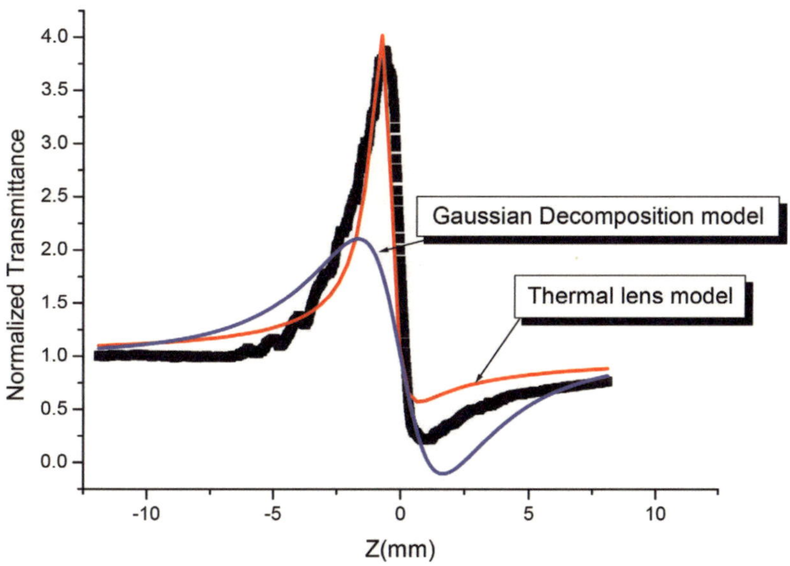

Figure 12.
Closed-aperture Z-scan effect of PdS quantum dots (radius 2.4 nm) irradiated with laser beam (λ = 488 nm and P_o = 6 mW); the data points (black) are fitted with the thermal lens model (red curve) and Gaussian decomposition model (blue curve).

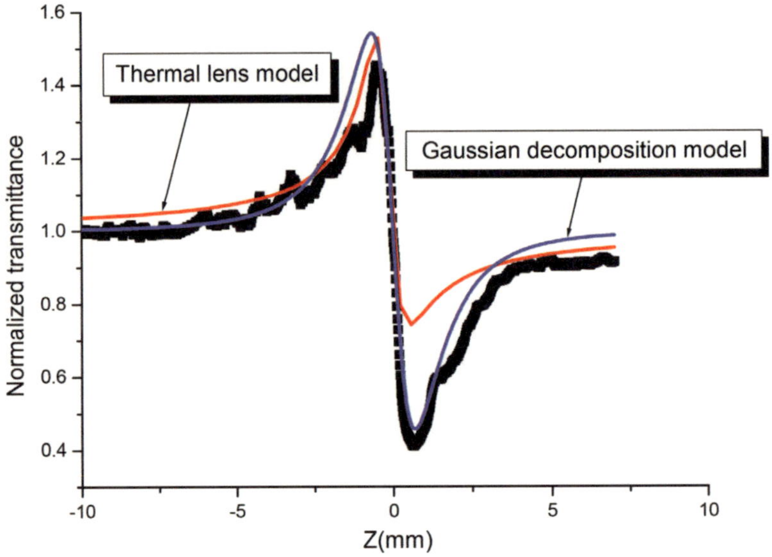

Figure 13.
Closed-aperture Z-scan effect of PdS quantum dots (radius 2.4 nm) irradiated with laser beam (λ = 633 nm and P_o = 6 mW); the data points (black) are fitted with the thermal lens model (red curve) and Gaussian decomposition model (blue curve).

5. Characterization of PbS QDs suspended in toluene

In this section, we will present—as an example—some nonlinear optical findings for samples of lead sulfide (PbS) quantum dots suspended in toluene. One sample of QDs has an average size (radius) of 2.4 nm and the other is with size 5.0 nm. It is very important to ensure that the sample is not contaminated with any impurities during QDs synthesis and storing or handling. The purity of the toluene quantum dot system was verified by NMR measurements where a small amount of the QD solution is dissolved in deuterated chloroform ($CdCl_3$). Therefore, any hydrogen in the solvent gets replaced by deuterium and only hydrogen will be originating from toluene. As shown in **Figure 14**, the peak at 0.00 ppm is for the TMS reference. The only peaks that appear are those by toluene that reveal the absence of any contamination that may influence both linear and nonlinear optical properties [25].

The samples were Z-scanned in both configurations OA and CA. Example of obtained results for QDs with radius 2.4 nm at three different excitation wavelengths with beam average power of 12 mW is shown in **Figure 15**. Similar general behavior is obtained with 5.0-nm QDs. **Figure 15** indicates clearly the dependence of third-order nonlinearity on excitation wavelength.

Figure 16 shows the closed-aperture Z-scan normalized transmittance for excitation wavelength of 633 nm, and 8-mW laser power for both the 2.4-nm and 5.0-nm QD radii. Similar transmittance curves are obtained for 488-nm and 514-nm laser excitation wavelengths. It is obvious from **Figure 16** that QD's size is a factor to be considered when quantifying third-order nonlinearity. The asymmetric curve is clearly indicative of nonlinear absorption, and the closed-aperture transmittance is affected by both nonlinear absorption and refraction. The suppression of the valley and enhancement of the peak of the transmittance curve imply that saturable absorption (SA) is the origin of nonlinear absorption effect. Whenever the excitation energy is much more than the optical bandgap or the excitation wavelength is near the peak of the absorption, saturable absorption becomes prominent due to high optical density.

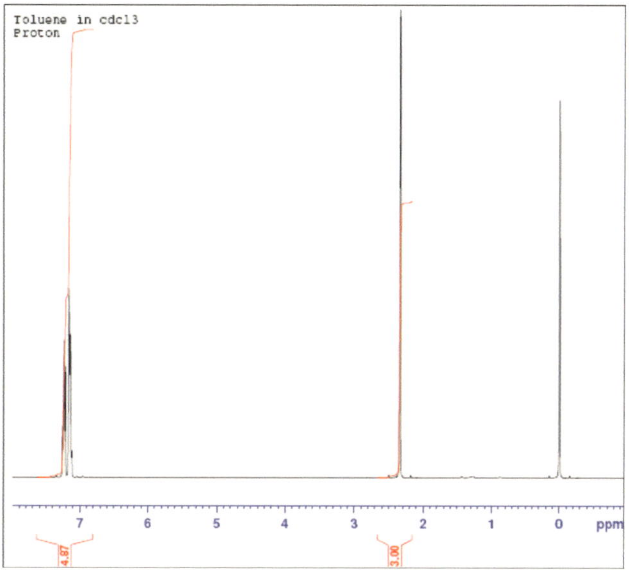

Figure 14.
NMR spectrum for sample of PbS QDs suspended in toluene [25].

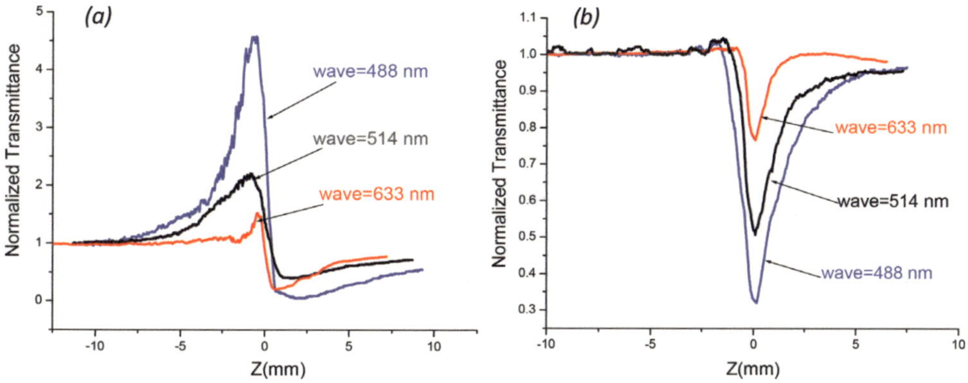

Figure 15.
Obtained normalized transmittances for QDs of radius 2.4 nm excited with laser beam power of 12 mW at three different wavelengths: (a) closed aperture configuration, and (b) open aperture configuration.

From **Figure 17**, we can not only tell the sign of the nonlinear refractive index (peak-valley = negative index) but also the origin of nonlinear refraction. Since CW lasers were used to investigate optical nonlinearity of samples, nonlinear refraction is due to thermal lensing which is a consequence of high linear absorption. Thermal lens model (solid line) was used to fit the data as presented in **Figure 17**. Moreover, it was suggested [16–19] that when the closed-aperture transmittance $\Delta T_{PV} > 1.7$, nonlinearity is due to thermal effect governing the nonlinear refraction behavior. This is clear in **Figure 17** where $\Delta T_{PV} \approx 4.5$.

Table 2 presents the calculated nonlinear refractive index n_2 for two different sizes of QDs excited with beam power of 8 mW, n_2 increases systematically with QD's sizes and the excitation wavelength. The variation of n_2 value with QDs size is illustrated in **Figure 18**. There is a clear trend of increase of n_2 with QD size for all wavelengths used.

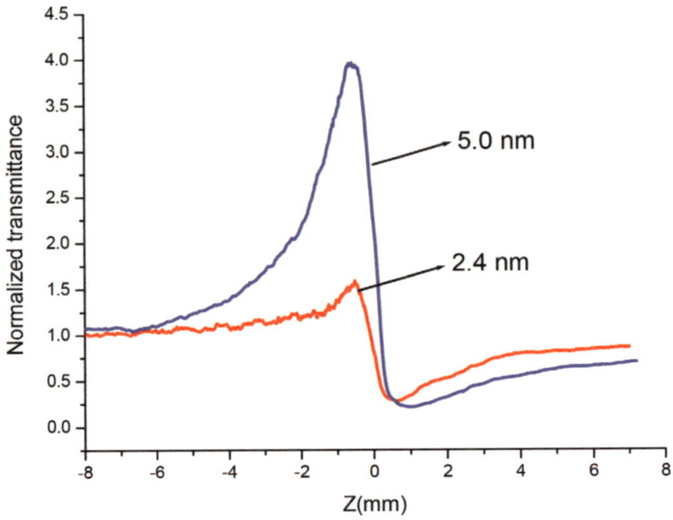

Figure 16.
Closed-aperture measurements for QDs with radius 2.4 nm and 5.0 nm samples scanned with beam of P_{in} = 8 mW and λ = 633 nm.

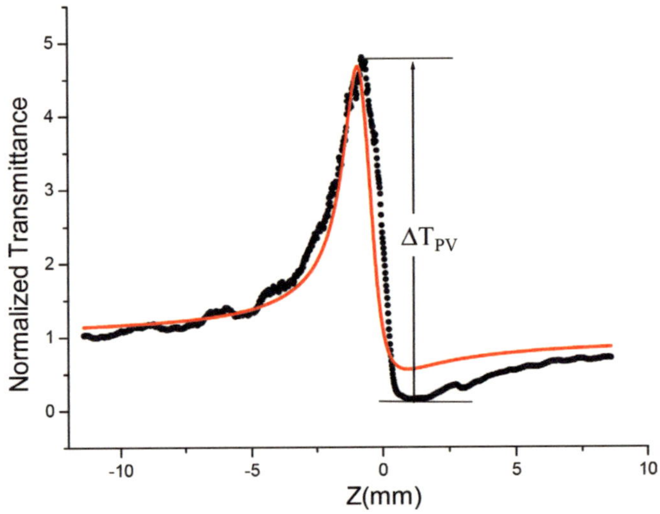

Figure 17.
Fitted closed-aperture measurements for 5-nm QD size samples scanned with beam of P_{in} = 8 mW and λ = 488 nm.

Quantum dot size (nm)	$n_2 \times 10^{-11}$ m²/W (λ = 488 nm)	$n_2 \times 10^{-11}$ m²/W (λ = 514 nm)	$n_2 \times 10^{-11}$ m²/W (λ = 633 nm)
2.4	3.501	3.742	4.631
5	5.383	6.975	8.406

Table 2.
Nonlinear index of refraction at different excitation wavelengths (P_{in} = 8 mW).

The nonlinear absorption coefficient β can be calculated from Eq. (17). T_{min} in this equation is obtained from the normalized open-aperture transmittance, as shown in **Figure 19**.

Table 3 shows the nonlinear absorption coefficient at different excitation wavelengths. For a given QD radius, there is a slight increase of the nonlinear absorption

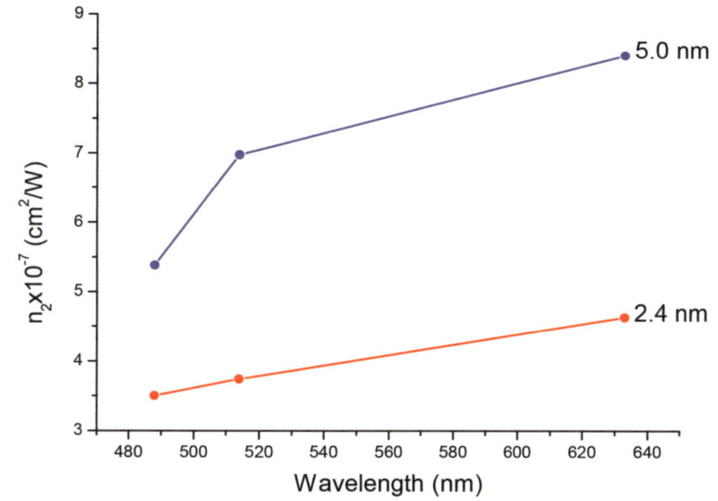

Figure 18.
Nonlinear index of refraction versus excitation wavelength (P_{in} = 8 mW) for different sizes of quantum dots.

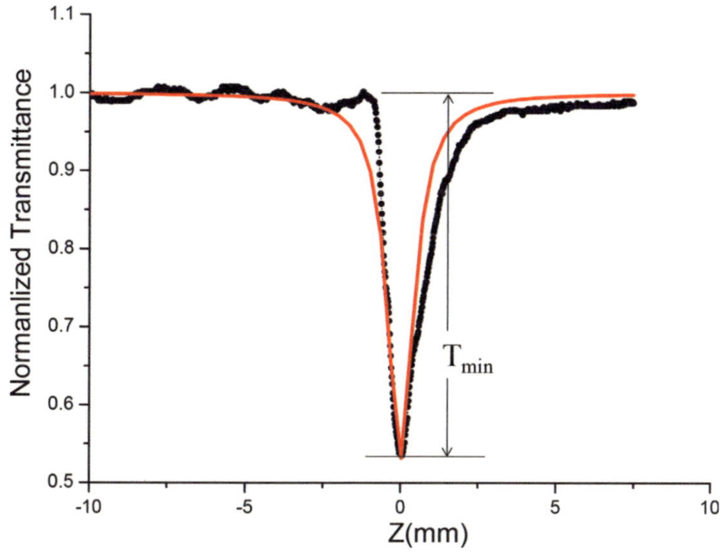

Figure 19.
Fitted open-aperture measurements for 5 nm QD size sample scanned with a beam of P_{in} = 8 mW and λ = 488 nm.

Quantum dot size (nm)	$\beta \times 10^{-5}$ m/W (λ = 488 nm)	$\beta \times 10^{-5}$ m/W (λ = 514 nm)	$\beta \times 10^{-5}$ m/W (λ = 633 nm)
2.4	8.676	17.12	21.530
5.0	**9.844**	**10.750**	**21.600**

Table 3.
Nonlinear absorption coefficient β at different excitation wavelengths (P_{in} = 8 mW).

coefficient with increasing wavelength. However, the change of the nonlinear absorption coefficient with size of QD is small.

Generally, the closed-aperture transmittance is affected by both the nonlinear refraction and nonlinear absorption. Therefore, the determination of n_2 is not very accurate from the closed-aperture scans alone. To obtain purely effective nonlinear refraction, we divide the normalized closed-aperture transmittance by the corresponding open-aperture scan. Then from proper fitting model (Eq. 24) of data, one can determine both nonlinear index of refraction and nonlinear absorption coefficients.

To calculate the third-order nonlinear susceptibility, the values of n_2 and β are first calculated, and then substituted in Eqns. (19) and (20). **Table 3** shows the calculated real and imaginary parts of the third-order nonlinear optical susceptibility $\chi^{(3)}$ and its absolute value $|\chi^{(3)}|$ for different values of n_2 and β. Investigating Eqs. (9) and (10), we observe that $\text{Re}\chi^{(3)}$ is proportional to n_2, and $\text{Im}\chi^{(3)}$ is proportional to β for a given excitation wavelength. This is the trend of variation we observe for both the real and imaginary parts of $\chi^{(3)}$.The figure of merit W (see Eq. 18) for different QD sizes is shown in the last column of **Table 4**. For small one-photon absorption α_0 in the near infrared region, the values W >>1 are ideal for applications in all-optical switching.

Table 5 presents the absolute value of the third-order nonlinear susceptibility $|\chi^{(3)}|$ for excitation wavelengths of 488 nm, 514.5 nm, and 632.8 nm at a constant laser beam power of 8 mW. Even though the real and imaginary parts of $\chi^{(3)}$ show no specific trend of variation with particle size, the absolute value $|\chi^{(3)}|$ increases systematically with particle size for the three wavelengths used in our Z-scan experiments. But, for the same QD size, it seems that $|\chi^{(3)}|$ is faintly enhanced as wavelength increases.

In fact, QDs act as excellent optical limiters, an example is presented in **Figure 7**. In order to study the optical limiting behavior of a sample using the Z-scan setup, one must place the sample in the post focal position and collect the transmitted light beam by closed-aperture configuration as illustrated in **Figure 20**.

Formation of diffraction fringes is presented in **Figure 21** for excitation power levels higher than 1 mW. Also, self-fanning dominates the transmittance spectrum. From the number of these fringes, one can estimate the change in the refractive index. Such interesting behavior ranks QDs media as very attractive active part of all optical switching devices.

In the above presented results, continuous-wave laser has been used and has resulted in thermally generated third-order nonlinearity due to deposition of heat. However, one way to overcome such effect is to use pulsed laser where electronic nonlinearity gets revealed. Moreover, one must avoid high power levels of

QD size (nm)	$n_2 \times 10^{-7}$ (cm^2/W)	$\beta \times 10^{-3}$ (cm/W)	Re $\chi^{(3)} \times 10^{-5}$ (esu)	Im $\chi^{(3)} \times 10^{-6}$ (esu)	$\vert\chi^{(3)}\vert \times 10^{-5}$ (esu)	W
2.4	3.501	8.676	2.025	1.949	2.034	4.345
5.0	5.389	9.844	3.116	2.211	3.124	3.807

Table 4.
Real and Imaginary parts of the third order nonlinear optical susceptibility for different QD sizes (λ = 488 nm, P_{in} = 8 mW).

QD size (nm)	$\vert\chi^{(3)}\vert \times 10^{-5}$ (esu), λ = 488 nm, P = 8 mW	$\vert\chi^{(3)}\vert \times 10^{-5}$ (esu), λ = 514.5 nm, P = 8 mW	$\vert\chi^{(3)}\vert \times 10^{-5}$ (esu), λ = 632.8 nm, P = 8 mW
2.4	2.034	2.202	2.750
5.0	3.124	4.042	4.902

Table 5.
The absolute value of the third-order nonlinear optical susceptibility for different QD sizes.

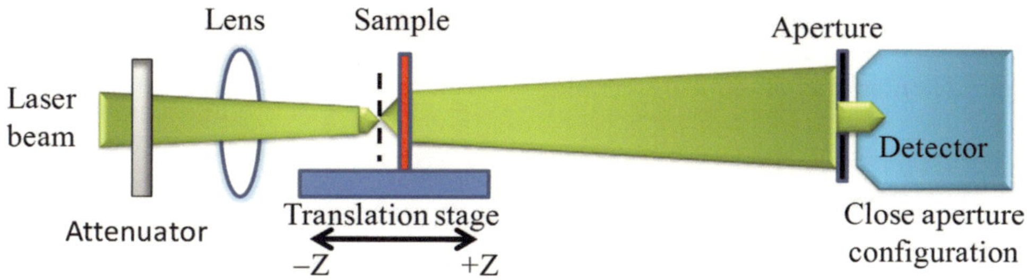

Figure 20.
Schematic of optical limiting investigation setup.

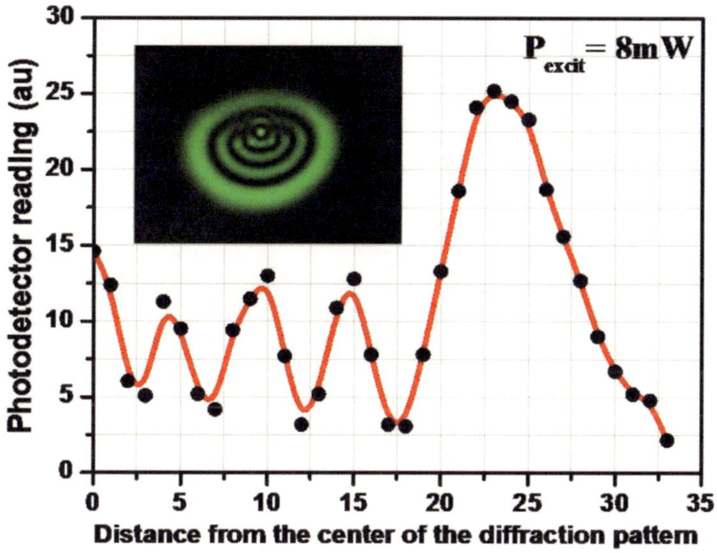

Figure 21.
A photograph of the diffraction rings due to nonlinear diffraction by QD sample of 2.4-nm radius Z-scanned with Ar^+ laser beam (P = 8 mW, and λ = 514 nm). The dots are the photodiode reading across the rings starting from the center of the diffraction pattern, the distance is in unit of mm [25].

excitation in order not excite higher orders of nonlinearity such as $\chi^{(5)}$ or exceeding the threshold damage of samples.

6. Conclusions

Quantum dots are interesting nanostructured materials. QDs are becoming an active part in many photonic devices. QDs possess unique optical and electronic properties. QD bandgap absorption edge is size dependent; as QD size is reduced, its bandgap energy gets larger. QDs possess large values of third-order nonlinear optical parameters such as nonlinear index of refraction n_2, nonlinear absorption coefficient β, and nonlinear susceptibility $\chi^{(3)}$. There are well-established models to explain and calculate third-order nonlinearity parameters obtained using different experimental techniques. Z-scan is considered an easy and practical technique used to investigate third-order nonlinearity for many classes of nonlinear media, among them are QDs. QD size, excitation wavelength, and incident power value are factors that need to be considered when investigating third-order nonlinearity. To separate electronic from thermal effects on quantified nonlinearity, one must use pulsed laser rather than continuous wave (CW). QD media are excellent active parts of optical limiters and all-optical switching devices.

Acknowledgements

The author would like to acknowledge the financial support provided by the University of Bahrain. Special thanks go to Professor Shawqi Al Dallal and Dr. Feryad Henari for their support and fruitful discussions.

Author details

Khalil Ebrahim Jasim
Department of Physics, University of Bahrain, Kingdom of Bahrain

*Address all correspondence to: kejasim@uob.edu.bh

References

[1] Jasim KE. Quantum dots solar cells. In: Kosyachenko LA, editor. Solar Cells —New Approaches and Reviews. InTech Open Access Publisher; 2015. pp. 303-331. DOI: 10.5772/59159. Chapter 11

[2] Brus LE. Electronic wavefunctions in semiconductor clusters. Journal of Physical Chemistry A. 1986;**90**: 2555-2560. DOI: 10.1021/j100403a003

[3] Boyed R. Nonlinear Optics. 3rd ed. Orlando: Academic Press Inc.; 2008. ISBN: 0123694701 9780123694706

[4] Zhao P. Ultrafast mechanisms of nonlinear refraction and two-photon photochromism [thesis]. Central Florida; 2016

[5] Rashidian M, Dorranian D. Investigation of optical limiting in nanometals. Reviews On Advanced Materials Science. 2015;**40**(12):110-126. ISSN 1605-8127

[6] Hassan QM, Al-Ahmad A, Al-Mudhaffer MF, Badran HA. Third-order optical nonlinearities and optical-limiting properties of phloxine B dye doped PMMA films investigated by Z-scan technique. Romanian Journal of Physics. 2012;**58**:962-969. ISSN 1221-146X

[7] Paul B. Investigations of nonlinear optical properties of certain organic photonic materials using Z-scan and DFWM techniques [thesis]. Cochin University of Science & Technology; 2004

[8] Saleh BEA, Teich MC. Fundamentals of Photonics. 2nd ed. John Wiley & Sons Inc.; 1991. DOI: 10.1002/0471213748. ISBN: 9780471213741

[9] Rekha RK, Ramalingam A. Nonlinear characterization and optical limiting effect of carmine dye. Indian Journal of Science and Technology. 2009;**2**(8): 29-31. ISSN: 0974-5645

[10] Sun XB, Wang XQ, Ren Q, Zhang GH, Yang HL, Feng L. Third-order nonlinear optical properties of bis (tetrabuty-lammonium)bis (4,5-dithiolato-1, 3-dithiole-2-thion) copper. Materials Research Bulletin. 2006;**41**: 177-182. DOI: 10.1016/j.materresbull. 2005.07.021

[11] Pramodini S. Third order optical nonlinearity and optical power limiting of organic materials under CW laser illumination [thesis]. Manipal University; 2015

[12] Krishnamurthy R, Alkondan R. Nonlinear characterization of mercurochrome dye for potential application in optical lmiting. Optical Application. 2010;**1**:187-196. ISSN 1899-7015

[13] Rashidian M, Dorranian D, Darani SA, Saghafi S, Ghoranneviss M. Nonlinear responses and optical limiting behaviour of Basic Violet 16 dye under CW laser illumination. Optik—International Journal for Light and Electron Optics. 2009;**120**(18): 1000-1006. DOI: 10.1016/j.ijleo.2008. 05.001

[14] Al-Saidi IA, Abdulkareem SA. Nonlinear optical properties and optical power limiting of Leishman dye using z-scan technique. Journal of Materials Science. Materials in Electronics. 2015; **26**(5):2713-2718. DOI: 10.1007/ s10854-015-2747-3

[15] Ganeev R, Ryasnyansky A, Redkorechev V, Fostiropoulos K, Priebe G, Usmanov T. Variations of nonlinear optical characteristics of C60 thin films at 532 nm. Optics Communications. 2003;**225**(1-3):131-139. DOI: 10.1016/j. optcom.2003.07.019

[16] Rao SV, Anusha PT, Prashant TS, Swain D, Tewari SP. Ultrafast nonlinear optical and optical limiting properties of phthalocyanine thin films studied using z-scan. Materials Sciences and Applications. 2011;**02**(05):299-306. DOI: 10.4236/msa.2011.25039

[17] Zongo S. Nonlinear optical properties of natural dyes based on optical resonance [thesis]. Western Cape; 2012

[18] Sheik-Bahae M, Said A, Wei T, Hagan D, Stryland EW. Sensitive measurement of optical nonlinearities using a single beam. IEEE Journal of Quantum Electronics. 2007;**26**(4): 760-769. **ISSN:** 0018-9197

[19] Sheik-Bahae M, Said AA, Van Stryland EW. High sensitivity, single-beam n_2 measurement. Optics Letters. 1989;**14**:955-957. DOI: 10.1364/OL.14.000955

[20] Sheik-Bahae M, Said AA, Wei TH, Hagan DJ, Van Stryland EW. Sensitive measurement of optical nonlinearities using a single beam. IEEE Journal of Quantum Electronics. 1990;**26**:760-769. DOI: 10.1109/3.53394

[21] Mathews SJ, Chaitanya Kumar S, Giribabu L, Vengopal Rao S. Nonlinear optical and optical limiting properties of phthalocyanines in solution and thin film of PMMA at 633 nm studies using a cw laser. Materials Letters. 2007;**61**: 4426-4431. DOI: 10.1016/j.matlet.2007.02.034

[22] Vinitha G, Ramalingam A. Spectral characteristics and nonlinear studies of methyl violet 2B dye in liquid and solid media. Laser Physics. 2008;**18**:37-42. DOI: 10.1134/S1054660X08010076

[23] Hou HW, Meng XR, Song YL, Fan YT, Zhu Y, Lu HJ, et al. Two-dimensional rhombohedral grid coordination polymers [M(bbbt)2 (NCS)2]n (M = Co, Mn, or Cd; bbbt = 1,1-(1,4- butanediyl) bis-1H-benzotriazole): Synthesis, crystal structures, and third- order nonlinear optical properties. Inorganic Chemistry. 2002;**41**(15):4068-4075. DOI: 10.1134/S1054660X08010076

[24] Cuppo FLS, Neto AMF, Gömeze SL, Muhoray PP. Thermal-lens model compared with the Sheik-Bahae formalism in interpreting Z-scan experiments on lyotropic liquid crystals. Journal of the Optical Society of America B. 2002;**19**:1342-1348. DOI: 10.1364/JOSAB.19.001342

[25] Khalil Ebrahim J, Al Dallal S. Fluorescence and nonlinear optics of PbS quantum dots dispersed in toluene. Journal of Experimental Physics. 2014;**1**: 19-28. ISSN 2384-4779

Chapter 4

Analysis of Pulsating White Dwarf Star Light Curves

Denis J. Sullivan

Abstract

Analysis techniques are presented for extracting the frequencies contained in the light curves of pulsating white dwarf stars. In several surface temperature regimes, these astronomical objects are unstable to gravity mode pulsations which result in brightness variations corresponding to the periods of the excited modes. There is a rich array of possible periods with values ranging from about 100 to 1000 seconds. Mode periods present in the light curve are detected by undertaking a Fourier analysis of the time series light curve; theoretical models of the star can be refined with this information. The Fourier analysis needs to take into account such things as finite length, data gaps and the presence of noise.

Keywords: white dwarf stars, pulsators, light curves, Fourier analysis, amplitude power spectra, prewhitening, false alarm probabilities

1. Introduction

The time series and their analyses covered here derive from those white dwarf stars that exhibit light variability as a consequence of their undergoing pulsation. In order to provide a context for this work it is appropriate to outline some key astrophysical details.

Many stars at particular evolutionary stages pulsate due to instabilities created by escaping radiation generated in the hot interiors interacting with partially ionized outer layers in the star. The pulsations cause surface brightness variations; recording and analysing the light curve of a pulsator enables investigators to determine internal properties of the star.

The pulsating white dwarfs are particularly productive targets for these investigations.

1.1 White dwarf stars

White dwarf stars [1] are the cooling remnants of about 98% of all stars. They are extremely dense as they have about the mass of the Sun compacted into a sphere about the size of the Earth. Typical densities are of the order of a million times that of water. Nuclear reactions have ceased and the inward pull of gravity in the object is offset by a pressure created by the quantum behavior of the (separate) dense electron gas.

This (degeneracy) pressure is independent of temperature, and hence provides permanent support down to the final “black dwarf” stage with temperatures approaching absolute zero. As the object cools the radiation output reduces in intensity and is centered around increasingly longer wavelengths. Eventually, the paucity of radiation emitted from the surface of this extremely cold object will render it invisible.

1.2 Pulsating white dwarf stars

During their long cooling sequence white dwarfs pass through several temperature regimes in which they are observed to pulsate [2]. Something like 80% of white dwarfs have a core consisting of a mixture of carbon-12 and oxygen-16 nuclei (plus the separate electron gas) overlaid by thin shells of helium and then hydrogen on top. Another significant group has only a thin helium surface layer overlaying the core.

At surface temperatures around 12,000 K for hydrogen atmosphere white dwarfs (25,000 K for the helium atmosphere types), the escaping radiation from the core interacts with a partially ionized subsurface layer, leading to the pulsations. The pulsations in most stars are called p-modes as they involve pressure variation propagation that results in radial material motion. This mechanism is the same as that which enables the propagation of sound in fluids.

Due to the extremely dense white dwarf material, the predicted periods for this type of oscillation are in the seconds domain. Although white dwarf p-modes have been searched for they have never been observed, probably due to their very low amplitudes (if they exist) combined with the difficulty of making suitable sub-second contiguous observations (even using 6 m class telescopes).

1.3 Gravity mode pulsations

The pulsations that are observed are nonradial gravity modes (g-modes). These involve largely angular movement of material in the star and have periods in the 100–1000 seconds domain [2]. They involve buoyancy variations in the stellar material; a simplified visual model of the mechanism involves an object floating on the surface of a fluid that will undergo vertical oscillations if it is pushed into the fluid. These oscillations only occur in a gravitational field, and in a spherical star they can only be nonradial.

The significant period difference between the two types of pulsation is directly connected to the larger forces involved in the compressibility (squeezing) of the dense white dwarf material versus the smaller buoyancy forces created by density differences.

The g-mode oscillations in white dwarfs were a serendipitous discovery in the late 1960s; theory predicts a large number of these g-modes with different periods. In the pulsators some of these modes are sufficiently exited such that their presence is detectable. The oscillations in effect result in temperature waves on the surface that produce observed variations at the mode frequencies in the light output of the star. An observed light curve of the white dwarf will therefore contain frequencies that yield information about the mode periods.

Detecting the different driven modes in a particular pulsating white dwarf allows comparison with predicted values from a detailed numerical model of the star [2]. Subsequent refinement of the model is then possible. In this way, detecting the various pulsation modes allows the observer to peer inside the otherwise hidden regions. Thus, one can measure such structure quantities as core chemical content and the masses and thicknesses of the overlying material shells. In some cases, the stellar rotation rate can be inferred.

1.4 Light curve time series

The raw materials for the time series are contiguous samples of the light output of a target star of interest. This is achieved using an instrument capable of fast photometry attached to a telescope [3, 4]. Ground-based telescopes dominate these observations, although space-based observations have become increasingly available.

Given the expected mode periods in white dwarfs, sampling rates of 10–20 seconds are typically employed. For the ground-based observations, gaps in the measurements due to the intrusion of clouds and the appearance of the Sun, as well as a signal random noise contribution, need to be accounted for in the analysis, along with the finite data length. Satellite observations can simplify these issues by providing uninterrupted data over extended intervals, although telescope aperture size is often a disadvantage.

The most useful information obtained from a light curve for modeling purposes is the periods of the detected modes. Hence, processing the time series in the optimum way in order to obtain a power spectrum is the primary goal. Since the amplitude of a driven pulsation results from a complicated nonlinear limiting process in the star, and is therefore more difficult to model, the detected amplitudes are primarily of use in differentiating between real and noise peaks. The phases of the different oscillations are in general not very useful, so a power spectrum is usually the ultimate analysis goal.

However, there is one case when the phase of a particular pulsation in an observation set is useful and that is when one is attempting to observe the frequency stability of the mode over an extended (e.g., multiyear) period. As long as one avoids cycle counting errors, tracking the measured phases over successive observation sessions can provide a sensitive test of the frequency stability of the mode.

The phase values for each observation set are best determined by least squares fitting sinusoids to the data. Then a predicted sinusoid maxima in a set is compared with the observed values from set to set to check for stability or any consistent changes. This is referred to as the Observed minus Calculated (O − C) method; for an example, see [5], but this will not be developed further here.

1.5 Time series examples

The observation of two types of pulsating white dwarf star obtained by the author will be used as time series examples in this article.

a. MY Aps (WD1) is a relatively bright example of the hydrogen atmosphere white dwarf pulsators.

b. QU Tel (WD2), a fainter target, is a member of the helium atmosphere pulsating white dwarf class.

Henceforth we will use the WD1 and WD2 labels to distinguish between the two objects rather than the astronomical variable names provided.

2. The raw light curve time series

The desirable time series form prior to undertaking a Fourier analysis consists of the light intensity excursions (i.e., modulation) about an average running mean of zero. The absolute detected light values are affected by such things as stellar distance, telescope aperture and detector efficiency; these quantities are not important in the analysis and certainly not in Fourier transform space.

A "typical" quality observation set is plotted in **Figure 1**. These data were obtained on the night of 12 July 2004 using a three-channel photometer [3] attached to a one meter telescope at the University of Canterbury Mt. John Observatory (UCMJO) in the South Island of NZ.

This photometer enables the measurement of the amount of light passing through three small identical circular apertures in the vicinity of the particular target object (WD2 here). Three identical photomultipliers operated in photon counting mode were employed in each "channel" to digitize the light intensities, and each plotted point in the graph corresponds to a 10 seconds integration of the counts.

The blue points correspond to the variable white dwarf, the green points to a nearby (brighter) constant comparison star, and the red points to the sky background light from a clear patch of sky. The comparison star data show the effect of observing through a changing amount of airmass as the object transits across the sky, along with the occasional intrusion of cloud, primarily around 15.5 hours.

The sky data show the effect of changing background light (on a moonless light in this case). Note that the white dwarf and sky points are plotted using the same scale. The target star here is rather faint and about half the light intensity is contributed by the sky background. For optimum measurements it important to use an aperture of an adequate size so that part of the image formed by the telescope optics (the point spread function) is not obscured.

Prior to processing the variable star data using Fourier techniques, one requires a time series of light curve modulation values, so conversion of the variable star data to fractional excursions about a running mean is required.

The first step, as detailed in the figure, is to divide the sky-corrected variable data by the sky-corrected comparison data and multiply by an appropriate constant

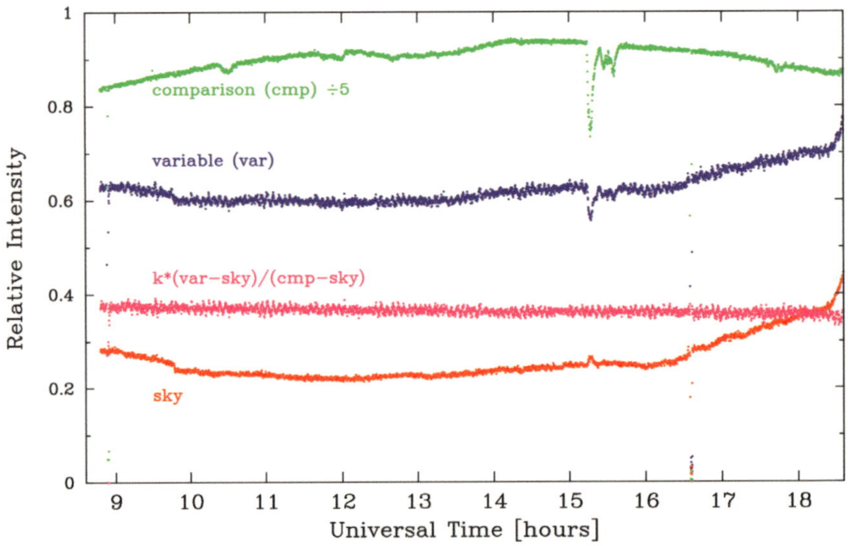

Figure 1.
Raw time series data obtained on the pulsating white dwarf WD2 on 12 July 2004 from the University of Canterbury Mt. John one meter telescope. The data were obtained using a photometer attached to the telescope that enabled the monitoring of three regions of the sky defined by small circular apertures [3]; identical photomultiplier tubes operated in photon counting mode measured the light intensities. The blue, green, and red plots correspond to the target white dwarf, a nearby (brighter) comparison star and a blank section of the sky, respectively. The mauve plot is the result of the indicated transform aimed to eliminate observing artifacts.

to yield a time-series curve with the correct mean value. This yields the mauve points plotted in **Figure 1**.

The second step is to convert light curve intensities to fractional deviation values about the running mean. A 3 hours section of the resulting data is presented in the middle panel (2) of **Figure 2**. A visual inspection of this plot reveals that along with a noise contribution there are oscillatory signals present.

Two intensity deviation measures are commonly utilized in this work. They are fractional or percent deviations, and millimodulation intensities (mmi) such that 10 mmi is equivalent to a 1% deviation. The latter measure is useful given the fact that real signal variations below 1% are detectable and often feature.

The light curve data displayed in **Figure 1** is of sufficiently high quality that almost no transformation steps beyond those listed above were required. However, useable time series light curves can be recovered from observations obtained in less than ideal photometric conditions by using a number of other techniques.

Such things as deleting obviously bad integrations and dividing the observations into separate groups to be "stiched together" later can be used. These manipulations are best accomplished interactively using a graphical display. As part of this process, the author developed a program [6] that enabled the interactive fitting of cubic splines to the light curves in order to remove "unwanted" low frequency variations. By implementing before and after Fourier transforms while manipulating the data, the user can directly see the impact of the changes and either accept or reject them.

The analysis will then of course be blind to the presence of possible low frequency signals, but ensuring that there is no impact on mode periods of $\sim$100 seconds or smaller will not impact the analysis. Some judgment is required in the whole process of extracting an optimum time series from the raw observations that can then be Fourier analysed.

The bottom panel (3) in **Figure 2** is a segment of observations made of WD2 in July 2003 using a charge coupled device (CCD) instrument camera attached to a 6.5 m aperture telescope in Chile (the Magellan Clay telescope). The CCD takes electronic images of the small field of view and in this case both the white dwarf and a nearby comparison star were captured, allowing reductions to proceed similarly to those described above.

The CCD exposure times were 20 seconds followed by a 10 seconds gap in order to allow the (serial) reading out of the individual CCD pixel charges and digitizing

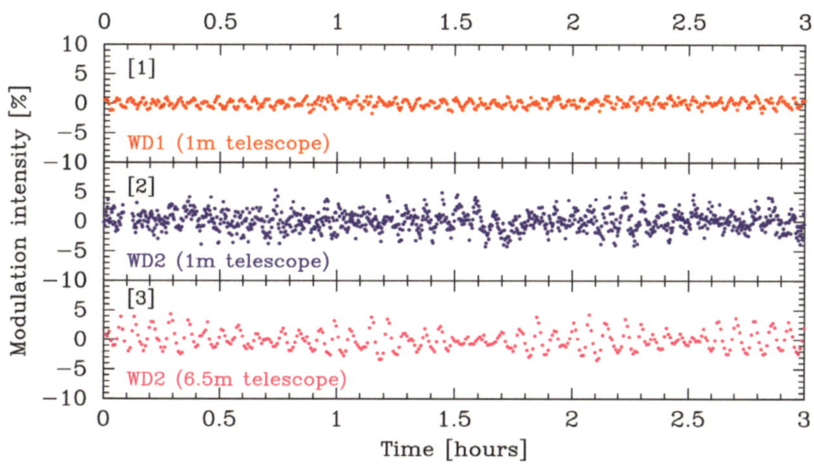

Figure 2.
Three hours sections of reduced light curve time series suitable for undertaking a Fourier analysis. The top panel (1) corresponds to WD1 and the middle (2) and bottom (3) panels to WD2. Note that the vertical scale is the same for the three panels and the major difference between the two WD2 plots is the telescope aperture size.

each amount. The shuttering of the CCD is required to prevent contamination from incoming photons while reading out. This means the system only received light from the stars for 2/3 of the observing time (see below). However, the much larger "light grasp" of this telescope provided significantly enhanced signal to noise measurements as can easily be seen by examining the plot.

It is clear that there are at least two closely spaced frequencies in the light curve by observing the *beating* (sinusoidal variation of the amplitudes). One can use both the period of the oscillations and the beating period to estimate the two frequencies and their separation in frequency space. We will leave this as an exercise and simply refer to the appropriate (amplitude) power spectrum in **Figure 3**.

The time series in the top panel (1) of **Figure 2** is a segment of the light curve of WD1 obtained in May 2013 using a CCD photometer attached to the Mt. John one meter telescope. This CCD photometer [4] is a frame transfer CCD that obviates the necessity of shuttering the device while reading out the previous exposure. This is accomplished by using a double sized CCD chip with one section masked off and used as a storage area. The analogue charges in the exposed CCD area can be rapidly moved to the storage area, and the next exposure proceed while the pixels in the storage area are being read out and digitized.

Each point in the plot corresponds to a 20 seconds exposure. The analogue charge shifting time is a small fraction of the exposure time so no shuttering is required. Minimal photon contamination occurs due to the rapid movement of the analogue charge.

It is clear from the time series plot that WD1 exhibits at least one pulsation mode with both a smaller period and a smaller amplitude compared with the WD2 oscillations.

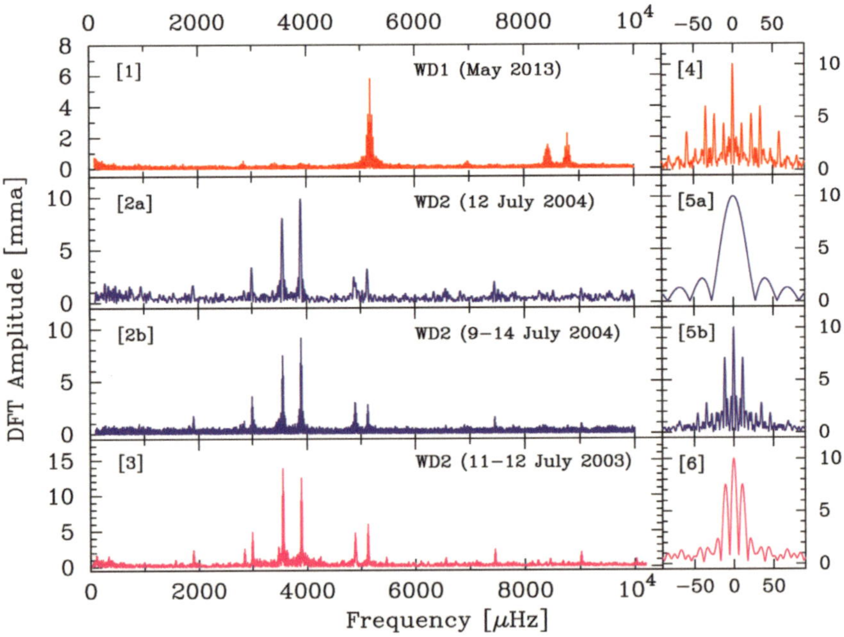

Figure 3.
Four amplitude power spectra (DFTs) resulting from a Fourier analysis of the full time series light curves corresponding to the segments depicted in **Figure 2**. *The observing window functions (see text) are presented alongside the DFTs. These functions show the impact of time series length and gaps on the DFT. Panel 1 results from observations of WD1 over a six night interval. Panels 2a and 2b correspond to observations of WD2, with 2a resulting from the light curve in* **Figures 1** *and* **2b** *the full six nights of observation. Panel 3 corresponds to two successive nights of the WD2 light curve plotted in the bottom panel of* **Figure 2**.

The introduction of CCD-based photometers (see for example, [4]) in the last few decades has made the acquisition of quality time series data a lot simpler; among other things all the star intensities and background measurements are made with effectively identical instruments. For sparsely populated observing fields the relevant intensities can be extracted from the images using synthetic aperture techniques [4], and if all images are recorded, the parameters can be varied offline to obtain optimum results.

3. Fourier analysis of the reduced light curves

Having suitably prepared the light curve time series, the next step is to perform a Fourier transform. As mentioned previously, the primary targets of the analysis are both the frequencies and amplitudes of the detected modes, with the phases mostly being of no interest. Therefore we require a power spectrum of the data calculated at specified frequency values.

It is much more enlightening to work in amplitude space, so the square root of a power spectrum is what is used in this work. Sometimes this result is called a periodogram. These amplitudes then match more closely the values one gets by least squares fitting sinusoids to the data; this will be discussed in the prewhitening section below.

Henceforth, we will simply describe the amplitude power spectrum used in this work as a discrete Fourier transform (DFT). Clearly we need to choose a set of frequencies at which to determine the DFT values.

It is informative to view each amplitude Fourier transform frequency point as a cross correlation of the data with a sine function defined by the chosen frequency and a variable phase factor. One obtains the correct amplitude value by varying the phase factor so as to obtain the maximum value. The standard algorithm does not actually do this, but instead performs the cross correlations separately with sine and cosine functions with fixed (zero) phases, and computes the square root of the sum of the squares of the two amplitudes thus determined. This is then the value of the (amplitude) DFT at the chosen frequency.

The complete DFT of interest is the set of amplitude values calculated for all the chosen frequencies.

This raises the interesting question of the choice of frequencies, the sampling rate in the time domain and the Nyquist frequency. For a given (uniform) sampling interval (δt) in the time domain, in principle there is frequency information in the DFT up to the Nyquist frequency defined by $f_N = 1/2\delta t$. This result follows from the fact that only two contiguous samples are required to define each sinusoid.

So for example, the choice of a light curve sampling value of 10 seconds yields a Nyquist frequency of $f_N = 1/20$ Hz $= 5 \times 10^4$ μHz. Note that with periods of 10^2–10^3 seconds of interest in the light curves, it is convenient to use the μHz =10^{-6} Hz frequency scale. Thus a typical mode period of 200 seconds corresponds to frequency of 5000 μHz.

Given current computational speeds it is more informative in the Fourier analysis procedures to use light curve sampling intervals that yield a Nyquist frequency significantly higher than the highest expected frequency of interest in the light curve.

So for the white dwarf pulsators with frequencies up to $\sim 10^4$ μHz of interest, 10 seconds ($f_N = 5 \times 10^4$ μHz) and 20 seconds ($f_N = 2.5 \times 10^4$ μHz) are suitable sampling times.

3.1 DFT versus FFT

A version of the very efficient Fast Fourier transform (FFT) algorithms was published by Cooley and Tukey in 1965 (see for example, [7]), and these computational methods became widely used in the following decades.

In effect the FFT algorithm trades increased algorithmic complexity and a certain lack of flexibility for (often vastly) improved computational times when calculating a Fourier transform. The algorithm is most efficient for a uniform sequence of samples with a number that is a power of 2, e.g., 2^M with M an integer; it removes many redundant DFT calculations by successively dividing the sequence into halves such that at the core one has a 2×2 transform; efficient computation steps are invoked to regain the full transform.

Compared with a direct calculation using the DFT method for the same number of frequency points, the computational speed difference between the two methods increases rapidly with sample number. For a sequence of length $N = 2^M$, the computing time for the FFT increases as $\sim N\log_2 N$ compared with $\sim N^2$ for the DFT.

However, in spite of this speed advantage, where practical it is advantageous to use the direct DFT for the Fourier analysis discussed here. And, the large advances in computing speeds have made this choice nearly always practical. The FFT algorithm requires uniform samples, and the frequency values and separations are defined by the number of time domain samples. This can be quite limiting and there is only one case in the analysis discussed here in which the author introduced use of the FFT (see conclusion section).

3.2 DFTs and the window function

DFTs of the time series light curves obtained for WD1 and WD2 are presented in the four panels in **Figure 3**, along with the associated window functions in the panels on the right. The vertical axes use the millimodulation amplitude (mma) scale, such that 10 mma corresponds to a signal amplitude of 1%, while the horizontal axes use the measure $\mu Hz = 10^{-6}$ Hz.

The dominant pulsation mode in WD1 (panel 1) has a peak amplitude of less than 1% at 6 mma. The two primary pulsation modes in WD2 (panel 3) have larger amplitudes ~13 mma. (note that the smaller amplitudes for WD2 displayed in panels 2a and 2b result from light contamination from a nearby faint star that could not be separated from the target image during the aperture photometry measurements made with the smaller aperture Mt. John telescope.)

The window function used here is an extended version of the window function that appears in the standard Fourier literature, which depicts the effect of a finite segment length on either the Fourier series or Fourier transform.

Using the simplest example, one can view a finite length time series of length T as being an infinite time series multiplied by a "top hat" function with a value 1 between $t = -T/2$ and $t = +T/2$ and otherwise zero. This will abruptly turn the signal on and off and therefore represent the actual time series of finite length T.

Fourier theory shows that multiplying two signals in the time domain corresponds to a convolution in the Fourier domain (and vice versa). The Fourier transform of a top hat is a sinc function of the form $\sin x/x$, so any frequency delta function peak in the spectrum will be expanded to the shape of a sinc function. In turn squaring these shapes to produce an amplitude power spectrum will result in peak shapes like the window function in panel 5a in **Figure 3**—as the time series here is that reduced from **Figure 1**, which is essentially a continuous section of 10 seconds samples over an ~9 hour interval.

What this result is telling us is the fact that in Fourier frequency space we are endeavoring to decompose the signal into sine and cosine functions that extend to $\pm\infty$, so we require a range of frequencies to interfere and turn the time series signal on and off.

For the real white dwarf time series that can have multiple and varied gaps, the shape of the window function is most readily determined by calculating values at all the data points of a (noiseless) sinusoid with some frequency (e.g., 5000 μHz here) and then computing the DFT of this function in the same way as for the data.

Only the shape and frequency scale is important, so the window functions for the four DFTs in **Figure 3** are plotted about a central value of zero. The expanded detail in the vicinity of peaks in a DFT can be inspected with the window function shape in mind.

For ground-based observing under generally good conditions, the basic frequency scale and shape of the symmetrical window function is dictated by two factors. The overall length of the time series determines the width of the central peak and the daily gaps dictate the accompanying satellite peaks. This is nicely depicted in the three window functions 5a, 6 and 5b in **Figure 3**.

The 5a window corresponds to a basically uninterrupted time series of length $\sim$10 hours = 3.6×10^4 seconds, so the width of the central peak is the inverse of this which is about 28 μHz. The window in panel 6 results from two approximately 9 hours observing sessions on successive nights, so the central peak is reduced accordingly with added alias peaks offset by the frequency $(\text{day})^{-1}$ = 11.6 μHz. Panel 5b continues this trend with observations over six successive nights along with daily gaps.

The panel 4 window function derives from observations over six successive nights so it has a narrow central peak related to the total interval of $\sim$5.5 days, the daily 11.6 μHz aliase peaks and additional peaks created by data gaps due to observing an alternative target.

4. Prewhitening the light curves

A very useful technique for identifying low amplitude frequencies in a light curve, especially when close to a larger amplitude one, is the procedure of prewhitening. To prewhiten a light curve of a given frequency one uses least squares fitting to determine the amplitude and phase of a given frequency, or more generally these quantities for a set of frequencies. One then computes a light curve based on these derived parameters and subtracts it point by point from the original light curve.

In this way, the frequency (or frequencies) are "removed" from the light curve. A DFT of this prewhitened light curve will reveal more details about the frequencies remaining.

Although least squares fitting can be used to determine the frequencies of interest, it is much simpler to use the frequency values determined from the DFT as then linear least squares fitting can be used to provide direct results. If the frequency values are also found using least squares then this unavoidably leads to a nonlinear least squares fitting problem. This is more complex and requires iterative techniques.

On the other hand, a sine function with a known frequency but an unknown phase can be expanded into the sum of sine and cosine functions, both with zero phases and unknown amplitudes. Such a combination of functions are suitable for linear least squares fitting.

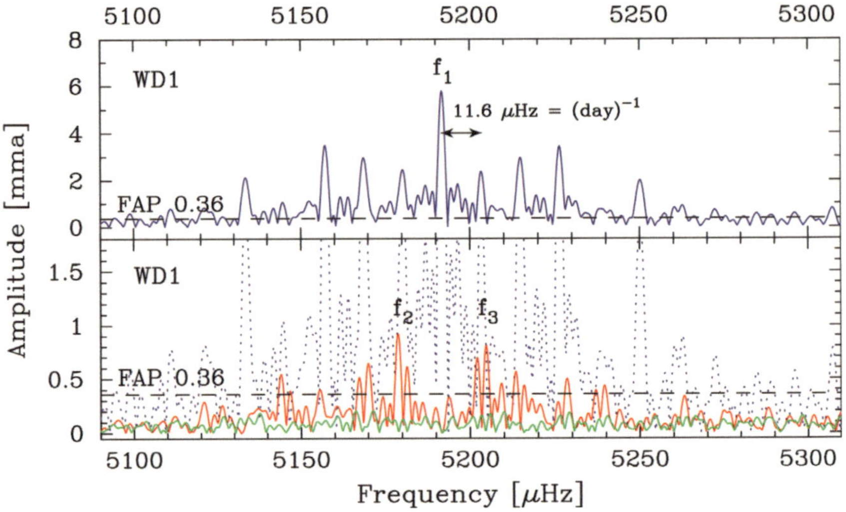

Figure 4.
An expanded section of the WD1 DFT presented in ***Figure 3*** *showing the structure around the dominant pulsation mode: $f_1 \sim 5190$ μHz, period ~192 seconds.*

Figure 4 demonstrates the utility of prewhitening using a section of the WD1 DFT displayed in the top panel of **Figure 3**. The top panel in **Figure 4** is an expanded version of the DFT around the dominant frequency f_1. The impact of the observing window function is clearly evident with the additional peaks surrounding the central one. This particular structure results from the daily breaks, together with additional peaks created by observing interruptions during the night due to the monitoring of another target star.

The red curve in the bottom panel of **Figure 4** shows the DFT of the light curve prewhitened by f_1. This reveals two frequencies, f_2 and f_3, of smaller amplitudes either side of the main peak. These peaks are also affected by the window function. The DFT with no prewhitening has been added to the plot as a dotted blue curve.

The green curve displaying essentially noise is a DFT of the light curve prewhitened by all three frequencies and clearly demonstrates the reality of the two smaller peaks. The horizontal dashed lines marked FAP 0.36 represent a significance level for the DFT peaks that is explained in the next section.

This analysis along with other WD1 observations will be published in Ref. [8].

5. Significance levels

As one examines peaks in a DFT that approach the noise level, it is important to adopt a quantitative criterion that facilitates differentiating between real and noise peaks. The False Alarm Probability (FAP) concept [9] provides such an intuitively straight forward approach.

Since we are dealing with statistical quantities, a particular FAP value (e.g., 0.001) will allow one to assert that there is a 1/1000 chance of a noise peak being as large as this value. So a peak exceeding this amount can be asserted to be real with a false alarm probability of 1/1000.

There are theoretical methods to estimate this number [9], but with the speed of modern computers a Monte Carlo approach is both enlightening and preferable (as with other uncertainty estimate requirements).

For the DFT FAP the method is as follows. First, the time series is prewhitened by the identified frequencies (in practice, the key requirement is to prewhiten by

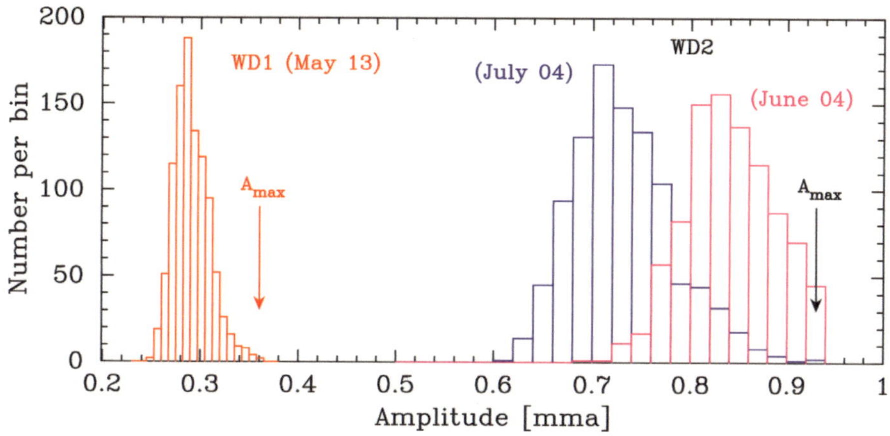

Figure 5.
Histograms of the maximum noise peaks obtained from the Monte Carlo noise analysis described in the text. The red histogram corresponds to the analysis of the WD1 DFT presented in panel 1 of ***Figure 3****. The blue histogram corresponds to the analysis of the WD2 DFT in panel 2b of* ***Figure 3****, and the mauve histogram to the DFT analysis of a similar multi-night WD2 observation set.*

the frequencies with larger amplitudes). Next, the data time stamps are randomly shuffled, a DFT calculated for this randomized time series and the largest noise peak selected. If this randomizing and DFT calculation process is repeated N times, with the largest DFT peak each time recorded, an ensemble of largest peaks is obtained. It can then be asserted that there is only a 1/N chance of random noise conspiring to produce a DFT peak as high as the largest peak in the ensemble.

It is instructive to plot histograms of these ensembles of largest peaks. Using a value for N of 1000, three examples are plotted in **Figure 5**. The red histogram corresponds to the May 2013 observing programme of WD1 yielding a value of 0.36 mma as the maximum value. Thus, there is a 0.001 probability of a random noise peak exceeding this peak value for the WD1 DFT. One can then assume with reasonable certainty that any peaks above this threshold are real with a false alarm probability of 0.001.

The blue and mauve histograms correspond to DFTs for WD2: the blue histogram relates to the DFT for WD2 plotted in panel 2b of **Figure 3** (see Ref. [6]), while the mauve histogram derives from multi-night photometry of WD2 obtained in June 2004 under similar conditions to that of July 2004. Although their shapes are different, both these histograms yield the same 0.001 FAP peak height of 0.93 mma.

Clearly the WD1 and WD2 histograms show that the WD1 data set is affected by significantly less noise; this is primarily the result of the white dwarf being a brighter target, but also in part due to the observations being obtained by a superior instrument. A comparison of the panels 1 and 2 in **Figure 2** qualitatively demonstrates a similar conclusion.

It should be emphasized that the FAP procedures outlined here relate only to random uncorrelated noise estimates. Nonrandom "noise" problems are mostly clearly identifiable on a case by case basis.

6. Conclusions

The analysis techniques described here relate specifically to those white dwarf pulsators that undergo coherent oscillations which are stable in frequency and phase over long periods. This is the class of isolated white dwarfs that exhibit gravity mode pulsations, which provide a potentially rich spectrum of modes that may be excited.

The amplitudes of some of the frequencies in a few pulsators are variable over intervals of weeks or months, with modes appearing and disappearing. However, the underlying frequency values indicating the mode structure remains stable. And, as mentioned above, it is this mode structure that is used to improve the theoretical models of the target white dwarf.

A class of pulsator that does not present stable coherent pulsations involves a pulsating white dwarf in a close binary configuration such that material is flowing from the companion on to the white dwarf. Although these are interesting objects, the Fourier techniques presented here are not as powerful [10].

The two pulsator examples used here are relatively low amplitude pulsators, but other white dwarf pulsators display large and nonsinusoidal light curve variations. This means the DFT can contain both harmonic and combination frequencies. The WD2 DFT shows some low amplitude combination frequencies [6], but none of these effects are present in the WD1 DFT.

The mode frequencies deduced from the WD2 time series observations enabled improved modeling of the white dwarf as well as an estimate of its rotation period of close to 2 hours [6]. Using the theory of "rotational frequency splitting," analysis of the WD1 data revealed a rotation period close to 12 hours [8]. In addition, when combined with single-site observations over several decades, at least one mode frequency was stable enough to enable the detection of an evolutionary cooling timescale [11].

6.1 Software techniques

The software tools described here have been developed in a Fortran 77/90 environment using the GFortran compiler provided by the GNU project. This personalized software has gradually evolved over a number of years, and the author has always found it more flexible to operate in a general programming environment rather than be limited by packaged products.

This certainly does not mean all software has been developed ab initio. Of considerable assistance has been access to the software routines both described and provided in the Numerical Recipes book series by Press et al. [12]. These authors in their last offering have also produced a version using the C++ programming language.

The Numerical Recipes routines, along with the comprehensive descriptions, provide an excellent base with which to develop appropriate software. The direct DFT described here is quite straight forward, but for other software developments the author has utilized Numerical Recipes routines (usually modified to suit) or ideas. Examples are: least squares fitting (both linear and nonlinear), randomizing routines, cubic spline fitting and FFT routines.

As mentioned previously, there was only one situation where the author decided the additional effort in using an FFT algorithm was worth it.

A multi-site multi-day observing campaign [6] on WD2 produced a total of 46,289 10 seconds samples over a 9 day interval. Calculating a false alarm probability at the 0.001 confidence level required 1000 DFT computations of a time series of this size. In order to exploit the power of the FFT, missing sample values were padded with zeros and the time series was extended to the size of 65,536 = 2^{16} by adding zeros. The overall computation time was vastly faster than using the direct DFT (which was also undertaken as a comparison and check).

Acknowledgements

The observational material discussed here was obtained at either the University of Canterbury Mt. John Observatory or the Magellan telescope facility in Chile, and the author thanks the relevant committees for the awards of observing time. The NZ Marsden Fund provided financial assistance for some of this work.

Author details

Denis J. Sullivan
School of Chemical and Physical Sciences, Victoria University of Wellington, New Zealand

*Address all correspondence to: denis.sullivan@vuw.ac.nz

References

[1] Tayler RJ. The Stars: Their Structure and Evolution. Cambridge University Press; 1994

[2] Winget DE, Kepler SO. Pulsating white dwarf stars and precision asteroseismology. Annual Review of Astronomy and Astrophysics. 2008;**46**

[3] Sullivan DJ. The New Zealand WET three channel photometer. Baltic Astronomy. 2000;7:100

[4] Chote P, Sullivan DJ, Brown R, Harrold ST, Winget DE, Chandler DW. Puoko-Nui: A flexible high-speed photometric system. Monthly Notices of the Royal Astronomical Society. 2014; **440**:1490

[5] Kepler SO et al. Measuring the evolution of the most stable optical clock G 117–B15A. The Astrophysical Journal. 2005;**634**:1311

[6] Sullivan DJ et al. Whole earth telescope observations of the hot helium atmosphere pulsating white dwarf EC 20058–5234. Monthly Notices of the Royal Astronomical Society. 2008;**387**: 137

[7] Fast Fourier Transform [Internet]. 2018. Available from: https//en. wikipedia.org/wiki/Fast [Accessed: November 22, 2018]

[8] Sullivan DJ et al. The hydrogen atmosphere pulsating white dwarf L 19–2 I: The 1995 WET campaign. Monthly Notices of the Royal Astronomical Society

[9] Scargle JD. Studies in astronomical time series analysis II: Statistical aspects of spectral analysis of unevenly spaced data. The Astrophysical Journal. 1982; **263**:835

[10] Chote P, Sullivan DJ. The post-outburst pulsations of the accreting white dwarf in the cataclysmic variable GW Librae. Monthly Notices of the Royal Astronomical Society. 2016;**458**: 1393

[11] Chote P, Sullivan DJ. The hydrogen atmosphere pulsating white dwarf L 19–2 II: Detection of an evolutionary timescale. Monthly Notices of the Royal Astronomical Society

[12] Press WH, Teukolsky SA, Vettering WT, Flannery BR. Numerical Recipes in Fortran. 2nd ed. Cambridge University Press; 1992

Chapter 5

Biotoxicological Monitoring of Organic Solvents in the Tunisian Footwear Industry

Imed Gargouri, Fatma Omrane and Moncef Khadhraoui

Abstract

Organic solvents (OS) are widely used in Tunisian footwear industry; however, there are no data related to employees' exposure. The objective of this study was therefore to adjust analytical methods in our laboratory for exposure assessment purposes. The predominant solvents are acetone, cyclohexane, hexane, methyl ethyl ketone, and toluene. Eighteen companies benefited from 55 airborne and 190 urine samples. Quantification of solvents and their metabolites was achieved by analytical methods that were adapted and validated in our laboratory. Airborne solvents were determined using gas chromatography (GC-FID). Urinary solvents or metabolites were measured either by GC or high-performance liquid chromatography (HPLC). Validation criteria were determined and used to judge the methods reliability. For airborne solvents, the concentrations exceeding the threshold limit value are mainly for hexane. For urines, the hippuric acid concentrations exceeded the biological limit value in semi-industrial process. Surprisingly, trans, trans-muconic acid was found in industrial and artisanal processes even though benzene was not among the used products. GC and HPLC methods have been adjusted, optimized, and effectively used to quantify OS and their metabolites in airborne and urine samples. Thus, a process of occupational risk assessment via a biotoxicological and airborne monitoring for solvents is now set.

Keywords: solvent exposure, chromatographic methods, risk assessment, indoor air, biomonitoring

1. Introduction

In the modern world, chemicals are integrated almost into every part of human life and activity. Their handling especially in huge quantities represents in some cases a health risk. In this context, organic solvents are regarded among the main chemical pollutants of the family called volatile organic compounds [1, 2–5] commonly used in the industrial sectors. Indeed, these products are widely employed in several fields and included in the composition of various products such as paints, inks, glues, pesticides, degreasers, and thinners. However, due to their readily volatilization, these solvents can be easily released into the atmosphere during manufacturing, storage, transportation, and application, which facilitate their inhalation by human and causing thus adverse health effects. In some cases, the inhaled chemical even undergoes biotransformations and may create more or less reactive intermediates leading to intoxication. Therefore, the assessment of human exposure to organic solvents in workplaces where they are handled in great quantities is of utmost importance for elucidation of human risks.

Among the chemical risks (CR) identified in Tunisian footwear manufacturing industry, organic solvents occupy by far the first place. However, despite the huge consumed quantities, exposure data of employees to these solvents in this area are almost absent. Actually, the prevention of occupational risks, especially for CR, is based on risk assessment following the procedures set by the regulations [4–8]. Nonetheless, the Tunisian health regulations [7, 8] have provided no requirement for entrepreneurs to conduct risk assessments via indoor air measurements in the workplaces or any biological exposure monitoring. We think that one of the reasons could be related to the lack of specialized laboratories in the field of metrology and health exposure assessment, since foreign companies engaged in such assessment are obliged to send their samples abroad.

To do so and overcome this shortcoming, the development of protocols and the validation of analytical methods to quantify the solvents and their metabolites in human fluids can partly solve the problem of exposure to OS. In the current case, the developed methods were used to evaluate the airborne content in terms of solvents such as acetone, cyclohexane, n-hexane, MEK, toluene, and trichloroethylene in shoe manufacturing companies in the city of Sfax. The detection and the quantification of their respective metabolites were also addressed.

Within this context, according to the guidelines, the most used technique is the gas chromatography (GC) coupled to a mass spectrometry detector [9–11]. The urinary metabolites of OS are also assessed by chromatographic methods; liquid chromatography (HPLC) [12, 13] and gas chromatography (GC) [14], each coupled to mass spectrometry, remain by far the most suitable methods for solvent and metabolite quantification. However, in our laboratory the GC is coupled with flame ionization FID detector. In this investigation, and as mentioned earlier, our main focus is to adapt the analytical method with our instrument and to validate GC-FID and HPLC methods to be used later on in routine analysis of OS and their respective urinary metabolites, in workplaces, such as shoe industry, with the ultimate aim to assess chemical exposures and health adverse impacts.

2. Material and methods

2.1 The shoe manufacturing process and sampling of the footwear companies

Actually, in Tunisia, the shoe industry, despite its mechanization, remains a labor industry where about 150 operations are required to make a pair of shoes. The shoe manufacturing steps are already presented in previous publications [15, 16].

The footwear manufacturing companies were identified and classified under three groups according to their manufacturing processes: industrial, semi-industrial, and artisanal. The classification is previously detailed [16] and briefly summarized in **Table 1**.

2.2 Selection of solvents and their metabolites to be quantified

Following the identification of solvents encompassed in the composition of products handled (glues, thinners, and strippers) in shoemaking conducted during the first half of 2008 in Sfax region, we were able to identify the most predominant solvents such as acetone, cyclohexane, hexane, methyl ethyl ketone, toluene, and

	Process	Industrial	Semi-industrial	Artisanal	Total
The discovered population (2005)	Companies	26	6	60	92
	Employees	751	48	350	1149
The selected sample (2008)	Companies	6	6	10	22
	Employees	122	48	60	230

Table 1.
Footwear companies.

a b

Figure 1
(a) Individual sampling of indoor air exposure. (b) Stationary sampling of the workplace's atmosphere.

trichlorethylene [4, 17, 18]. They were subjected to an airborne quantification in addition to their metabolites which were measured in urine worker samples [19]. Although benzene was not identified in the composition of the products, for security reasons, it was systematically investigated via its metabolite trans, trans-muconic acid in urine [7, 8, 20].

2.3 Sampling and analytical methods

The solvent exposure assessment was performed along 15 weeks, based on some former studies [5, 21, 22] and via several steps that are detailed in our previous publication [16].

To measure personal exposure, an air sampling holder was fixed near the respiratory zone of each volunteer worker (**Figure 1**) [23–27]. The sampling method and equipments have already been described [16]. These samples were taken in the middle of the week on Wednesdays or Thursdays depending on the type of the company's manufacturing process and installation locations (**Table 2**).

Result comparisons of the analyzed samples were carried out in reference to average values of exposure (*TLV*) calculated over a reference period of 8 hours/day and 39 hours/week (**Table 7**) [7, 8, 17, 28, 29].

In order to quantify solvent metabolites, 190 employees among 230 involved in shoe manufacturing, belonging to 22 companies, have benefited from urinary sampling. These samples were taken by the weekends and at the end of the work shifts:

Process manufacturing	Installation location	Leave weekly	Airborne sampling	Urinary sampling
Industrial	Industrial zone	Saturday (afternoon) and Sunday	Wednesday (Morning)	Friday (afternoon)
Semi-industrial				
Artisanal	The Medina (old town)	Sunday (afternoon) and Monday	Thursday (morning)	Saturday (afternoon)

NB.: Semi-industrial companies: some in the Medina and other industrial areas.

Table 2.
The moments of interventions in the companies.

Companies	Class	Identified	Planned	Realized	Airborne measurements	Urinary sampling
Industrial	1	26	6	4	17	105
Semi-industrial	2	6	6	5	23	60
Artisanal	3	60	10	10	15	25
	Total	92	22	18	55	190

Table 3.
Sample of participating companies to airborne and biomonitoring interventions.

on Friday or Saturday afternoon (**Tables 2** and **3**) [30–33]. These urine samples were performed with reference to the biological limit values (**Table 7**) [16, 34–36].

2.3.1 Toxicological analysis protocols

2.3.1.1 Dosing the target solvents in the indoor air

After desorption in 5 ml of carbon disulfide (Fluka®, Ref. 84710), the activated charcoal was analyzed by gas chromatography coupled with flame ionization detector (GC-FID) using an external calibration mode (**Tables 4** and **5**, **Figure 2**): (i) SHI MADZU® chromatograph, GC-9A, and (ii) capillary column Hewlett-Packard® HP-5MS (L = 60 m, inner diameter = 0.25 mm, film

Solvent		Acetone (C_3H_6O)	n-Hexane (C_6H_{14})	MEK (C_4H_8O)	Cyclohexane (C_6H_{12})	Trichlorethylene (C_2HC_{13})	Toluene (C_7H_8)	Benzene (C_6H_6)
Desorption	Solvent	Carbon disulfide						
	Amount (ml)	5 ml for validation area and 1.5 ml for the control area						
	Mode	Manual agitation						
	Stirring (min)	30 min at relatively low temperature						
	Materials	Volumetric flasks, rocking stirring, 10 ml vials of glass, micropipette						
Volatilization temperature		56.3°C	68.7°C	79.6°C	80.7°C	87.3°C	110.6°C	88°C
Method		GC-FID (Gas chromatography with flame ionization detection)						
External standard		Acetone (C_3H_6O)	n-Hexane (C_6H_{14})	MEK (C_4H_8O)	Cyclohexane (C_6H_{12})	Trichlorethylene (C_2HC_{13})	Toluene (C_7H_8)	Benzene (C_6H_6)

MEK: methyl ethyl ketone.

Table 4.
Preparation and analytical targets solvents "activated charcoal tube: 200/800" [26, 35–37, 44].

Solvent	Acetone	n-Hexane	MEK	Cyclohexane	Trichlorethylene	Toluene
Retention time (min)	6.2	7.7	8.3	9.4	10.9	15.3
Area peak	882337	3552515	20758	2107460	892254	2678766
Concentration (mg/l)	14.10	11.78	14.37	13.89	26.16	15.48

Table 5.
Retention time (min) concentration (mg/l) and surface area of peaks of the various solvents.

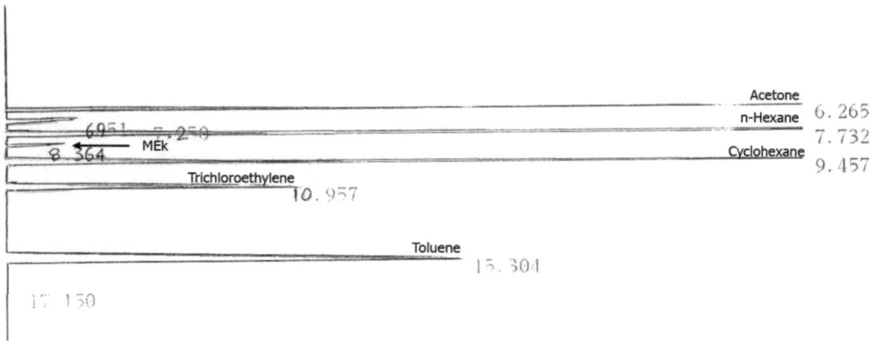

Figure 2.
Chromatogram of a standard injected in the same analysis conditions as those of the samples.

thickness = 0.25 μm). To optimize this method, different conditions were tested such as oven temperature, split/splitless injections, injected volume, and column type and length [37–39].

2.3.1.2 Dosing the metabolites of target solvents in urine

Depending on the nature of the metabolites of solvents, the following techniques were used for their quantification (**Figure 3** and **Table 6**) [37, 44–49]:

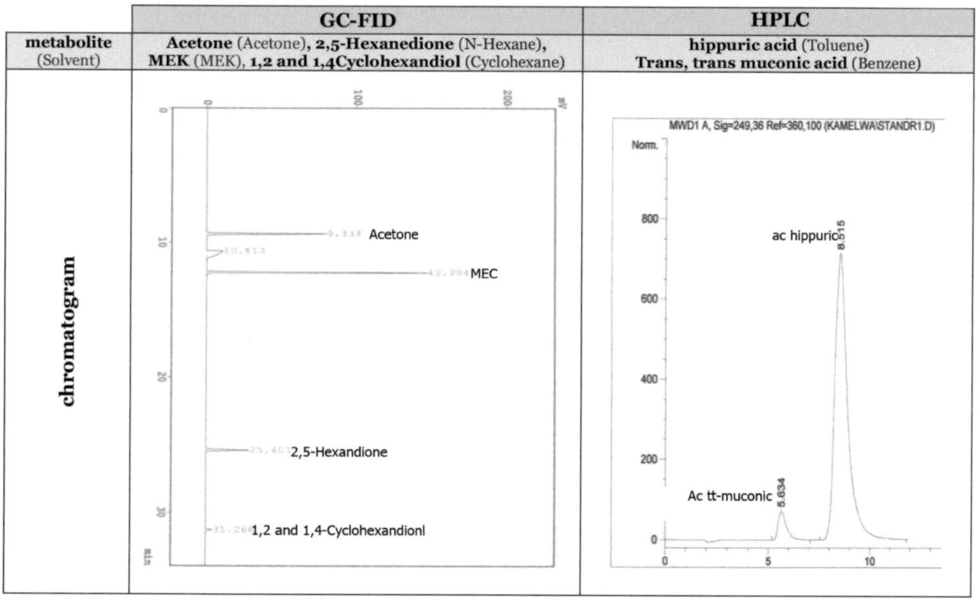

Figure 3.
Chromatograms of two standard injected under the same conditions as analytical samples.

	GC-FID			HPLC		
Metabolites	**Acetone**	**2,5-Hexanedione**	**MEK**	**1,2 and 1,4-Cyclohexandiol**	**Trans, trans-muconic acid**	**Hippuric acid**
Retention time (min)	9.333	12.204	25.401	31.266	5.6	8.5
Area peak	379899	762130	176371	29836	2062.5	31054.3
Concentration (mg/l)	3507.1	3614	957	174.4	13.5	1214

Table 6.
Retention time (min) concentration (mg/l) and peak area of the different metabolites.

- Either by GC-FID:
    - SHI MADZU® chromatograph GC-17A
    - Capillary column Hewlett-Packard® HP-1 (L = 60 m, inner diameter = 0.25 mm, film thickness = 1 μm)
- Or by HPLC:
    - Chromatograph agilent 1100 series
    - Column shim-pack CLC-ODS (n) 15 cm

As mentioned before, HPLC and GC conditions were optimized, and validation criteria (repeatability, detection limit, linearity, and recovery test) were evaluated.

2.4 Biometrological exposure limit values

In the absence of Tunisian exposure guideline values for the workplace air quality and biological samples [7, 8], we referred to the French, American, and/or German values. Then, we adopted the most severe among them as reference values (**Table 7**) [17, 28–31, 50, 51].

More specifically, the air concentrations were compared to threshold limit value-time-weighted average (TLV-TWA) for solvents.

For the biological exposure indices (BEI) related to metabolites (**Table 7**) [16, 34–36], they were as follows: acetone for acetone, 2,5-hexanedione for hexane, 1,2-cyclohexanediol for cyclohexane, methyl ethyl ketone for methyl ethyl ketone, hippuric acid for toluene, and trans, trans-muconic acid for benzene [32, 52, 53].

3. Results

3.1 Airborne solvent quantification

All conducted measurements in different companies showed that the exposure to solvents varies from one process to another and from one station to another [16]. Direct results of airborne concentration measurements of the solvents are published in our previous publication [16]. Average airborne concentrations of hexane are particularly high especially with overruns of the TLV in upper worker, sole maker, and finishing positions in all processes except for the artisanal type 2. For the other

DOI: http://dx.doi.org/10.5772/intechopen.86295

Solvent	No. CAS[1]	la France (UE[2]) VME[6]		USA (ACGIH[3]) TLV-TWA[7]		Germany (MAK[4])		Our study (Tunisia) Adopted VM		Biological Exposure Index (BEI) "Urine"	la France (EU) VGF[9] (1997)	USA (ACGIH) BEI[10]	Germany (DFG[5]) BAT[11]	Our study (Tunisia) IBE adopted
		ppm[8]	mg.m⁻³	ppm	mg.m⁻³	ppm	mg.m⁻³	ppm	mg.m⁻³					
Acetone	67-64-1	500	1210	500	-	500	1200	**500**	**1200**	Acetone	100 mg/l	50 mg/l	80 mg/l	**50 mg/l**
cyclohexane	110-82-7	200	700	300	-	200	700	**200**	**700**	1,2-cyclohexanediol	-	-	170 mg/g creat	**170 mg/g creat**
n-Hexane	110-54-3	20	72	50	-	50	180	**20**	**72**	2,5 hexanedione	-	0.4 mg/l	-	**0.4 mg/l**
Methyl ethyl ketone (MEK, 2-butanone)	78-93-3	200	600	200	-	200	600	**200**	**600**	Methyl ethyl ketone	-	2 mg/l	5 mg/l	**2 mg/l**
Toluene	108-88-3	50	192	50	-	50	190	**50**	**190**	hippuric acid	2500 mg/g creat	1600 mg/g creat	-	**1600 mg/g creat**
trichlorethylene	79-01-6	75	405	50	-	-	-	**50**	**-**	trichloroacetic acid	100 mg/g creat	15 mg/l (Proposal 2007)	A	**15 mg/l**
Benzene	71-43-2	1	3.25	0.5	-	1	3.2	**0.5**	**1.6**	trans, trans-muconic	5 mg/l	0.5 mg/g creat	B	**0.5 mg/g creat**

(A) [Trichlorethylene] in air (ml/m³)	[Trichloroacetic acid] in urine (mg/l)
10	20
20	40
30	60
50	100

(B) [Benzene] in air (ml/m³)	[Trans, trans-muconic acid] in urine (mg/l)
0.3	-
0.6	1.6
0.9	-
1.0	**2.0**
2.0	3.0
4.0	5.0
6.0	7.0

[1]*CAS, chemical abstract service;* [2]*EU, European Union;* [3]*ACGIH, American Conference of Governmental Industrial Hygienists;* [4]*MAK, maximum-Arbeitsplatz konzentration;* [5]*DFG, Deutsche Forchungs-Gemeinschaft;* [6]*VME, exposure average value calculated with respect to a reference period of 8 hours;* [7]*TLV-TWA, time-weighted average (weighted average values over 8 hours per day and 40 hours per week);* [8]*ppm, parts per million by volume of air;* [9]*VGF, value guide French;* [10]*EIB, biological exposure indices;* [11]*BAT, Bioloischer Arbeitsstoff-Toleranz-Wert (biological values tolerated in the workplace).*

Table 7.
Values of airborne and biological exposure limits for solvents studied [9, 10, 13, 17, 25, 27, 28].

quantified solvents (acetone, cyclohexane, methyl ethyl ketone, and toluene), they were relatively high without exceeding the TLV.

3.2 Dosage of urinary metabolites

We limited ourselves to dosing urine samples collected during the first period: a stage where companies are on average activity (from 28 May to 28 July 2008).

3.2.1 Dosage of metabolites by HPLC

Table 8 summarizes the main data urinary dosages of two solvent metabolites: the hippuric acid and trans, trans-muconic acid that are the respective biomarkers

Process	Working post	Operator	n	Hippuric acid (mg/g creatinine)		Trans, trans-muconic acid (mg/g creatinine)	
				Range	Average	Range	Average
Industrial *"1 company"* 13 men and 9 women	Quilting	**Man**	4	124.3–222.3	159.1	0.8–3.5	1.7
	Upper worker	*Man*	2	80.8–242.2	161.5	0.7–1.6	1.2
		Women	5	33.9–352.5	190.9	0.9–4.4	2.4
		Total	7	33.9–352.5	182.5	0.7–4.4	2.1
	Sole maker	**Man**	5	57.0–359.7	192.7	0.2–6.0	3.1
	Finish	*Man*	2	67.7–138.0	102.8	1.7–1.9	1.8
		Women	3	71.5–381.7	194.7	1.5–5.5	2.8
		Total	5	67.7–381.7	158.0	1.5–5.5	2.4
	Serigraph	**Women**	1	—	180.3	—	1.6
Semi-industrial *"3 business"* 25 men and 8 women	Chopped off	**Man**	2	207.3–6768.1	3487.7	0.1–0.7	0.4
	Quilting	**Man**	1	—	749.3	—	0.0
	Upper worker	*Man*	9	380.4–9322.3	3112.7	00–03	0.1
		Women	3	25.5–5358.6	3523.9	00–02	0.1
		Total	12	25.5–9322.3	3215.5	0.0–0.3	0.1
	Sole maker	**Man**	11	546.7–7277.2	2349.4	0.0–0.4	0.1
	Finish	*Man*	2	2178.0–2609.5	2393.8	0.0–0.4	0.2
		Women	5	0.0–2823.9	1117.6	From 0.0 to 0.0	0.0
		Total	7	0.0–2823.9	1482.2	0.0–0.4	0.1
Artisanal *"6 companies"* 13 men	Upper worker	*Man*	4	295.8–1775.0	903.4	0.0–2.2	1.5
	Sole maker	*Man*	5	57.5–4582.7	1086.3	0.1–2.9	1.6
	Upper worker/sole maker	*Man*	4	440.2–4784.5	3292.5	0.0–0.2	0.1

Table 8.
Assays of urinary hippuric acid and trans, trans-muconic acid according to workstation.

of toluene and benzene. The parameters are the number of urine samples collected (n) per workstation and sex, the arithmetic mean of urinary concentrations of metabolites, and the range of concentrations.

The hippuric acid was particularly high with average exceeding the TLV for certain employees in the semi-industrial process for the majority of workstations, including the cutting. We noted particularly high values of the toluene in type 2 artisans, while the TLV was not exceeded in the industrial process.

The trans, trans-muconic acid was highlighted in the industrial process and artisanal type 1 with average ranging from 1.2 to 3.1 mg/g creatinine, far exceeding the TLV.

3.2.2 Dosage of metabolites by GC

Data on urinary dosage of metabolites of the four solvents, acetone, 2,5-hexandione, methyl ethyl ketone, and 1,2-cyclohexandiol, which are the respective biomarkers of acetone, hexane, cyclohexane, and MEK were not usable due to a technical problem in GC.

4. Comments and discussion

4.1 Reviews and bias

We had to suffer from some delay in achieving our airborne sampling and therefore an impact on the quality of our data because of different difficulties. In large part, it is due to the heaviness of administrative procedures to follow in Tunisia for the acquisition of scientific equipments. We have waited for over a year to have the air sampling pumps (May 2006, making contact with the supplier until June 2007, receipt of the order). Meanwhile, we adjusted the analytical analysis protocols for solvents and their metabolites. It required bibliographical research and repeated tests in the laboratory as it was not a directed technology transfer (North-South) [22, 39–43].

We were able to quantify the airborne samples with activated charcoal tubes after their storage at 4°C, and that was achieved within a short time. We did not have that opportunity for urine samples, and we had to freeze them since we were neither (and technicians) capable of performing these dosages nor owning sufficient material (one GC). This GC had also some technical problems, and we have not been able to use these results since conservation methods have not been respected. Only the results of the metabolites made by HPLC (hippuric acid and trans, trans-muconic acid) were analyzed.

4.2 Airborne sampling equipments

This material is of great interest to develop this type of action and toxicological metrology in the Sfax region. Indeed, this is the first time that there has been the acquisition of active sampling materials in the field of occupational health in Tunisia. Moreover, it is also the first opportunity on the establishment of a structured approach in occupational and environmental toxicology through GEET laboratory (previously known as 3E) in a new theme "impact of hazardous substances on environment and human health."

4.3 Biometrological measurements

For a dozen years, the use of solvents is undergoing a revolution, because of occupational risk prevention constraints but mostly because of regulatory requirements for environmental protection. These regulatory changes led to changes in the nature of the applied solvents and how to use them [17, 29, 54]. Meanwhile the number of exposed workers is growing in Tunisia. This is confirmed on the international level (in France): the SUMER 2003 survey showed that the number of employees exposed to solvents has increased since 1994 from 12.2 to 14.7% mainly in the chemical industry [2].

Thus, risk assessment studies in various sectors using solvents were started, but the shoe manufacturing sector remained unexplored. This is the case of the study of Poirot and Hubert-Pelle [17] that evaluated exposures to solvents with airborne samples in various industrial activities but not in the manufacture of glue.

If these airborne sampling and biotoxicological analyses were made for the first time in the sector of footwear manufacturing in Sfax, they included a sample of companies from the three manufacturing processes (industrial, semi-industrial, and artisanal); this was preceded by a preliminary risk assessment along with job tasks examinations and an inventory of the handled products [5, 6]. This risk assessment could be improved due to the experience we have gained and with the best knowledge of the sector and the risk prioritization in it.

The results of airborne sampling have confirmed that the existing gaps between the different shoe companies and different workstations were generally those that were indirectly estimated by the workstations observations.

All the conducted sampling in the companies indicates that employees' exposure to organic solvents is quite variable depending on the performed job task.

The hippuric acid was highlighted in the urines of some employees with exceedances of the limit value; however, the TLV of toluene has not been exceeded in the companies. This could be explained by a dermal exposure, especially since we have not recorded the use of gloves by workers when they handle preparations used in shoe manufacturing (glues, thinners, strippers).

We recorded the presence of trans, trans-muconic acid in urine analyses with exceedances of the biological limit value set at 0.5 mg/g creatinine (**Table** 7) for some employees in industrial or artisanal shoe companies. In contrast, the inventory of the handled products in the processes did not show the presence of benzene or unleaded gasoline. So, this is due either to the contamination of the used solvents by impurities or an environmental contamination by car exhausts (unleaded gasoline 95) especially since the majority of the employees in these shoe companies use two-wheeled vehicles (bicycle or motorcycle).

On the other hand, the analyses of some samples of the products that are handled in the manufacturing of shoes are highly recommended. Furthermore, it also recommended to start an environmental study to check the air quality in the major thoroughfares in Sfax, a city known for its pollution, and to make urine sampling in order to explore the benzene metabolite for people who are exposed to car exhaust (such as traffic wardens, auto mechanics) and unexposed ones.

5. Conclusion

GC and HPLC methods have been adjusted, optimized, and effectively used for the determination of OS and their metabolites in airborne and urine samples of solvent manipulators. The exploitation of these indicators had necessitated the use of new techniques for occupational surveillance for the first time in the region. Thus, a process of occupational risk assessment via a biotoxicological and airborne monitoring for solvent exposures is now set.

This study allowed us to provide information on chronic exposure to solvents in the shoe industry and to establish an initial observation on solvent exposure profiles in this sector. However, we know that exposure to solvents is not constant over time and varies according to the task performed and the utilized process. Therefore, the investigation on exposure needs not only the average exposure in comparison with TLV but also to identify the polluting phases in order to determine the short-term exposure.

In the footwear manufacturing sector, following the identification phase of the used solvents and the highlighting of overruns in airborne concentrations of some hazardous products, we plan to focus on the carcinogenic characteristic of certain preparations and especially their potential toxicity for reproduction, since we noted the increase of female presence in the shoe manufacturing sector and some cases of couple sterility.

Acknowledgements

The authors thank the University of Sfax for financial support.

Conflict of interest

Authors have declared that no conflict of interest exists.

Author details

Imed Gargouri*, Fatma Omrane and Moncef Khadhraoui
Laboratory "Environmental Engineering and EcoTechnology" LR16ES19, National School of Engineering, University of Sfax, Tunisia

*Address all correspondence to: imed.gargouri@fmsf.rnu.tn

References

[1] Dubeau M. Vapors and gases. In: Occupational Hygiene. 4th ed. Quebec, Canada: Griffen Inc Clay; 1985. pp. 415-523. ISBN: 2-920210-34-3

[2] Testud F, Lambert-Chhum R, Bellemin B, Descotes J. Exposition toxique professionnelle chez la femme enceinte-1 re partie: Principes de l'évaluation du risque individuel. Journal de Gynécologie Obstétrique et Biologie de la Reproduction. 2001;**30**: 776-779

[3] Wilson MP, Hammond SK, Nicas M, Hubbard AE. Worker exposure to volatile organic compounds in the vehicle repair industry. Journal of Occupational and Environmental Hygiene. 2007;**4**:301-310

[4] Dor F, Bonvallot N. Hazard identification: A step in the evaluation of health risks deepen. Environment, Health and Risk. 2005;**6**:279-287

[5] Gonzalez M, Velten M, Cantineau A. Solvent exposure assessment in an epidemiological study of 249 employees. Archives des Maladies Professionnelles de Médecine du Travail. 1999;**60**: 432-434

[6] INERIS (Institute National Industrial Environment and Risks). Health Risk Assessment Guide. INERIS; 2003. 152p. Available from: http://chimie.ineris.fr/fr/lespdf/guide_ERS.pdf [Accessed: October 17, 2008]

[7] JORT. Law No. 94-28 of 21 February 1994 on damage compensation scheme for industrial accidents and occupational diseases. JORT 1994;**15**: 308-318

[8] JORT. Law No. 95-56 of 28 June 1995 on the compensation scheme for damages of occupational accidents and occupational diseases in the public sector. JORT 1995;**53**:1419-1424

[9] NIOSH Manual of Analytical Methods (NMAM), Hrsg. Volatile Organic Compounds, C1 to C10, Canister Method. Available from: https://www.cdc.gov/niosh/nmam/pdf/3900.pdf

[10] Huard M. Analytical Method: Determination of Volatile Organic Compounds in the Air by Mass Spectrometry. Available from: https://www.irsst.qc.ca/media/documents/PubIRSST/MA-369-EN.pdf

[11] INRS. MétroPol: COV M-338. 2019. Available from: http://www.inrs.fr/publications/bdd/metropol/fiche.html?refINRS=METROPOL_338

[12] INRS. Benzène (71-43-2)/Acide trans, trans muconique urinaire—Biotox. Available from: http://www.inrs.fr/publications/bdd/biotox/dosage.html?refINRS=Dosage_104

[13] INRS. Toluène (108-88-3)/Acide hippurique urinaire—Biotox. Available from: http://www.inrs.fr/publications/bdd/biotox/dosage.html?refINRS=Dosage_99

[14] INRS. n-Hexane (110-54-3)/2,5-Hexanedione totale urinaire—Biotox. Available from: http://www.inrs.fr/publications/bdd/biotox/dosage.html?refINRS=Dosage_342

[15] Gargouri I, Khadhraoui M, Elleuch B. What are the health risks of occupational exposure to adhesive in the shoe industry? In: Rudawska A, editor. Adhesives—Applications and Properties. London: InTech; 2016. ISBN: 978-953-51-2783-3 978-953-51-2784-0

[16] Gargouri I, Khadhraoui M. Metrology organic solvents in the shoes industry to Sfax city (Tunisia). In: Yilmaz F, editor. Paint and Coatings Industry. London: IntechOpen;

2019. ISBN: 978-1-78985-161-8 978-1-78985-162-5

[17] Poirot P, Hubert G. Pelle exposure profiles to solvents and compared to the values short-term limits. ND 2235-200-05. Notebook Notes. 2005;**200**:83-93

[18] French Society of Occupational Medicine. Medical surveillance of employees exposed to toxic substances for fetal development. Archives des Maladies Professionnelles et de l'Environnement. 2005;**66**:165-176

[19] Hertsenberg S, Brouwer D, Lurvink M, Rubingh C, Rijnders E, Tielemans E. Quantitative self-assessment of exposure to solvents among shoe repair men. The Annals of Occupational Hygiene. 2007;**51**:45-51

[20] Cocco P, Tocco MG, Ibba A, Scano L, Ennas MG, Plant C, et al. Trans, trans-muconic acid excretion in relation to environmental exposure to benzene. International Archives of Occupational and Environmental Health. 2003;**76**: 456-460

[21] Armstrong TW, Caldwell DJ, Verma DK. A proposed methodology for setting occupational exposure limits for hydrocarbon solvents. Journal of Occupational and Environmental Hygiene. 2005;**2**:600-607

[22] Mckee HR, Medeiros AM, Daughtrey WC. A proposed methodology for setting occupational exposure limits for hydrocarbon solvents. Journal of Occupational and Environmental Hygiene. 2005;**2**:524-542

[23] Hervé-Bazin B. The evaluation guide of exposure to toxic risk in sampling workplace atmosphere. ND 1730-135-89. Notebook Notes. 1989;**135**: 265-288

[24] INRS. Database "Metropol". Available from: www.inrs.f [Accessed: March 20, 2006]

[25] IRSST (Robert-Sauvé Research Institute in Occupational Health and Safety)—Operations Directorate. Sampling Guide for Air Contaminants in the Workplace. 8th ed. IRSST, Canada; 2005. T-06: 191p. Available from: www.irsst.qc.ca [Accessed: August 15, 2006]

[26] IRSST. Record Substances Sampling Guide of Air Contaminants in the Workplace. Available from: www.irsst.qc.ca/fr/_listersst.html [Accessed: March 20, 2006]

[27] SKC Inc. Air Sample Pumps, Calibrators & Accessories. Sorbent Sample Tubes and Media Collection. Available from: http://www.skcinc.com/ [Accessed: March 15, 2006]

[28] Journal official. Decree No. 2007–1539 of 26 October 2007 setting binding occupational exposure limit values for certain chemicals and amending the Labor Code (Second Part: Decrees in Council of State). JORF, 28 October 2007: Text 13 of 52 (NOR: TST0760893D)

[29] JOUE. Directive 2006/15/EC of 7 February 2006 establishing a second list of indicative lists occupational exposure under Directive 98/24/EC of the Council and amending Directives 91/322/EC and 2000/39/EC. OJEU 9.2.2006: L36-L39 (FR)

[30] Pillière FF. Conso Biotox: Biotoxicologique Guide for Doctors. INRS (France) in September 2007, ED 791: 252p. Available from: www.inrs.fr [Accessed: September 15, 2007]

[31] Triplet J. Occupational exposure limits to hazardous substances. ACGIH values (USA) and the MAK commission (Germany). ND 2114-176-99. Notebook Notes. 1999;**176**:59-90

[32] Truchon G. Biomonitoring Guide: Collecting and Interpreting Results.

6th ed. Canada: IRSST; 2004. T-03: 92p. Available from: www.irsst.qc.ca [Accessed: May 19, 2005]

[33] William W, Ernest L, Christiani DC. Biomarker research in occupational health. Journal of Occupational and Environmental Medicine. 2005;**47**: 145-153

[34] Grandjean P, Budtz-Jørgensen E. Total imprecision of exposure biomarkers: Implication for calculating exposure limits. American Journal of Industrial Medicine. 2007;**50**:712-719

[35] Schneider O, Brondeau MT. Biological exposure indices. ND 2245-202-06. Notebook Notes. 2006; **202**:49-65

[36] Schulz C, Angerer J, Ewers U, Kolossa-Gehring M. The german human biomonitoring commission. International Journal of Hygiene and Environmental Health. 2007;**210**: 373-382

[37] Boeckelmann I, Pfister EA. Influence of occupational exposure to organic solvent mixtures in contrast sensitivity in printers. Journal of Occupational and Environmental Medicine. 2003;**45**:25-33

[38] Delinsky AD, Bruckner JV, Bartlett MG. A review of analytical methods for the determination of trichlorethylene and its major metabolites chloral hydrate, trichloroacetic acid and dichloroacetic acid. Biomedical Chromatography. 2005;**19**:617-639

[39] Eller PM, Cassinelli ME. NIOSH Manual of Analytical Methods (NMAM). 4th ed. USA: CDC; 1994. Available from: www.cdc.gov/niosh/nmam [Accessed: August 15, 2006]

[40] Kawai T, Miyama Y, Horiguchi S, Sakamoto K, Zhang ZW, Higashikawa K, et al. Possible metabolic interaction entre hexane solvents other co-exposed at sub-occupational exposure limit levels. International Archives of Occupational and Environmental Health. 2000;**73**:449-456

[41] Pergiago JF, Cardona A, Marhuenda D, Roel J, Villanueva M, Marti J, et al. Biological monitoring of occupational exposure to n-hexane by exhaled air analysis and urinalysis. International Archives of Occupational and Environmental Health. 1993;**65**:275-278

[42] Soulage C, Perrin D, Berenguer P, Pequignot JM. Sub-chronic exposure to toluene at 40 ppm alters the monoamine biosynthesis rate in discrete brain areas. Toxicology. 2004;**196**:21-30

[43] Calafat AM, Kklenyik Z, Caudill SP, Ashley DL. Urinary levels of trichloacetic acid, a disinfection by-product in chlorinated drinking water, in a human reference population. Environmental Health Perspectives. 2003;**111**:151-154

[44] Johns DO, Dills RL, Morgan MS. Evaluation of dynamic headspace with gas chromatography/mass spectrometry for the determination of 1,1,1-trichloroethane, trichloroethanol, and trichloroacetic acid in biological samples. Journal of Chromatography B. 2005;**817**:255-261

[45] Lof A, Hjelm EW, Colmsjo A, Lundmark BO, Norström A, Sato A. Toxicokinetics of toluene and urinary excretion of hippuric acid after-human exposure to toluene. British Journal of Industrial Medicine. 1993;**50**: 55-59

[46] O'Donnell GE, Juska A, Geyer R, Faiz M, Stalder S. Analysis of trichloroacetic acid in the urine of workers occupationally exposed to trichlorethylene by capillary gas chromatography. Journal of Chromatography A. 1995;**709**:313-317

[47] Perico A, Cassinelli C, Brugnone F, Bavazzano P, Perbellini L. Biological

monitoring of occupational exposure to cyclohexane by urinary 1,2- and 1,4-cyclohexanediol determination. International Archives of Occupational and Environmental Health. 1999;**72**: 115-120

[48] Takeuchi A, Kawai T, Zhang ZW, Miyama Y, Sakamoto K, Higashikawa K, et al. Toluene, xylenes and xylene isomers in urine as biological indicators of low-level exposure to each solvent: A comparative study. International Archives of Occupational and Environmental Health. 2002;**75**:387-393

[49] Takeuchi A, Kawai T, Zhang ZW, Miyama Y, Sakamoto K, Higashikawa K, et al. Comparison of n-hexane unchanged in alveolar air and 2,5-hexanedione in urine for the biological monitoring of n-hexane exposure in human volonteers. International Archives of Occupational and Environmental Health. 2004;**77**:264-270

[50] Courtois B. Occupational exposure limits to chemicals in France. National Institute for Research and Security [INRS] (France) in December 2007, ED 984: 19p

[51] Poirot P, Subra I, Baudin V, Hery M, Chouanière D, Vincent R. Determining medium-term exposure profile of house painters. Notebook Notes. 2000;**179**: 5-13. ND2125-179-00

[52] Kawai T, Yasugi T, Mizunuma K, Horiguchi SI, Ikeda M. Comparative evaluation of blood and urine analysis as tool for biological monitoring of n-hexane and toluene. International Archives of Occupational and Environmental Health. 1993;**65**: S123-S1S6

[53] Tkaukiainen A, Vehmas T, Rantata K, Numinen M, Martikainen R, Taskinen H. Results of laboratory tests common in solvent-exposed workers. International Archives of Occupational and Environmental Health. 2004;**77**: 39-46

[54] JORF. Decree No. 2008–602 of 25 June 2008 on recycling and waste from the textile apparel products, new house shoes or clothes for households. JORF No 149 of 27 June 2008 (NOR: DEVP0772293D)